COLOR TELEVISION TROUBLESHOOTING

EDWARD BANNON

RESTON PUBLISHING COMPANY, INC.
A Prentice-Hall Company
Reston, Virginia

Technical Drawings by Robert Mosher

Library of Congress Cataloging in Publication Data

Bannon, Edward
    Color television troubleshooting.

    Includes index.
    1. Color television — Repairing. I. Title.
TK6670.H48   621.3888'7   76-8518
ISBN 0-87909-130-4

© 1976 by Reston Publishing Company, Inc.
A Prentice-Hall Company
Reston, Virginia 22090

10 9 8 7 6 5 4 3 2 1

Printed in the United States of America

Table of Contents

Preface

With the rapid advance in electronic technology, combined with the marked shift in academic emphasis, particularly in the consumer-electronics area of the junior-college curriculum, the need has become apparent for a color-television guidebook that provides high-level motivation for students in the upper-class division. This text is a teaching tool that presents state-of-the-art color-television servicing information in a functional format that minimizes the prerequisites for effective application. It permits the upper-class student to test, diagnose, and repair many of the common defects and malfunctions that occur in solid-state color-television receivers.

The text treatment assumes that the student has completed a course in radio and television theory, and that he has obtained a working knowledge of the basic electronic servicing instruments. Although no prior instruction in color-TV circuit action is prerequisite, it is desirable that the student carry a concurrent course in color-TV theory. This text is profusely illustrated for minimization of the learner's burden and for optimizing retentivity. General principles of circuit function and malfunction are stressed throughout, with emphasis on logical reasoning from trouble symptom(s) to component defects. Test and measurement techniques are explained and illustrated for each basic subdivision of troubleshooting technology. The final chapter covers specialized digital-logic test instruments in application to troubleshooting digital-control and display systems in modern color-TV receivers.

This guidebook is the outcome of extensive teaching experience on the part of both the author and his fellow instructors. They have also contributed numerous constructive criticisms and suggestions. It is appropriate that this text be dedicated as a teaching tool to the instructors and students of our junior colleges, vocational schools, and technical institutes.

Edward Bannon

Basic Troubleshooting Approach

1 . 1   OPERATING AND MAINTENANCE CONTROLS

Color television troubleshooting starts with an analysis of the picture and sound trouble symptoms. As an illustration, a color receiver might display a picture that has lost color synchronization, or a picture that has lost both black-and-white and color synchronization, as shown in Fig. 1–1. In these examples, the sound output is almost always normal. Note that the loss of black-and-white and/or color sync may be caused by a defective component in the synchronizing circuitry of the receiver, or it may be caused by a simple misadjustment of the horizontal locking (hold) control. Therefore, it is essential to have a good understanding of the operating and maintenance controls that are provided in various color-TV receivers. Typical operating controls are shown in Fig. 1–2. These include a channel selector, fine-tuning control, volume control with off-on switch, ultra high frequency (UHF) tuning control, tint, color, and brightness controls, horizontal and vertical hold controls, contrast control, and tone control.

There is no sharp dividing line between operating controls and maintenance controls. As an illustration, the horizontal and vertical hold controls depicted in Fig. 1–2 are included with the maintenance-control group in some receivers. Operating controls

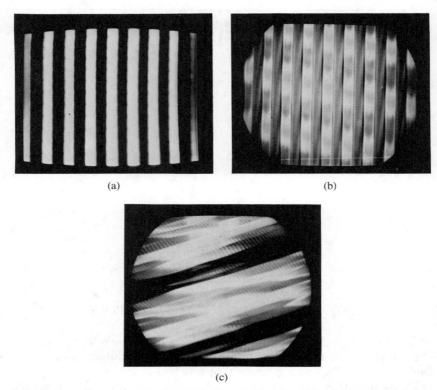

(a)   (b)

(c)

Figure 1–1. Examples of color picture trouble symptoms: **(a)** Normal chroma bar pattern; **(b)** Loss of color sync; **(c)** Loss of both color and black-and-white sync.

are located on the front or side (sometimes the top) of the receiver, and are readily accessible to the viewer. On the other hand, maintenance controls are usually mounted on the back of the receiver, or sometimes inside the cabinet. Typical maintenance controls include convergence, height, automatic gain control (AGC), color killer, and peaking controls, as exemplified in Fig. 1–3. Note that the "dots" control is not provided in most color receivers; this control produces a dot pattern on the screen that is used during adjustment of the convergence controls. This topic is detailed subsequently. Note in passing that a tone control is included with the maintenance controls in Fig. 1–3, whereas the tone control is grouped with the operating controls in the example of Fig. 1–2.

Each operating and maintenance control is associated with a

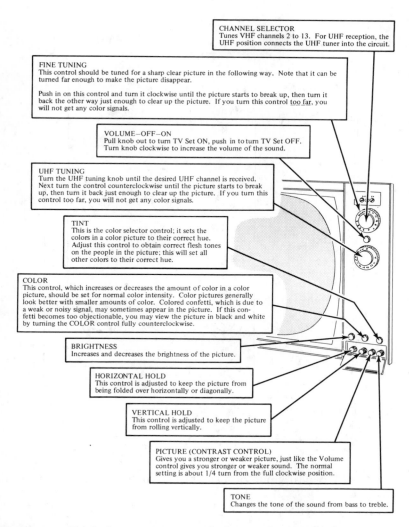

CHANNEL SELECTOR
Tunes VHF channels 2 to 13. For UHF reception, the UHF position connects the UHF tuner into the circuit.

FINE TUNING
This control should be tuned for a sharp clear picture in the following way. Note that it can be turned far enough to make the picture disappear.

Push in on this control and turn it clockwise until the picture starts to break up, then turn it back the other way just enough to clear up the picture. If you turn this control too far, you will not get any color signals.

VOLUME–OFF–ON
Pull knob out to turn TV Set ON, push in to turn TV Set OFF. Turn knob clockwise to increase the volume of the sound.

UHF TUNING
Turn the UHF tuning knob until the desired UHF channel is received. Next turn the control counterclockwise until the picture starts to break up, then turn it back just enough to clear up the picture. If you turn this control too far, you will not get any color signals.

TINT
This is the color selector control; it sets the colors in a color picture to their correct hue. Adjust this control to obtain correct flesh tones on the people in the picture; this will set all other colors to their correct hue.

COLOR
This control, which increases or decreases the amount of color in a color picture, should be set for normal color intensity. Color pictures generally look better with smaller amounts of color. Colored confetti, which is due to a weak or noisy signal, may sometimes appear in the picture. If this confetti becomes too objectionable, you may view the picture in black and white by turning the COLOR control fully counterclockwise.

BRIGHTNESS
Increases and decreases the brightness of the picture.

HORIZONTAL HOLD
This control is adjusted to keep the picture from being folded over horizontally or diagonally.

VERTICAL HOLD
This control is adjusted to keep the picture from rolling vertically.

PICTURE (CONTRAST CONTROL)
Gives you a stronger or weaker picture, just like the Volume control gives you stronger or weaker sound. The normal setting is about 1/4 turn from the full clockwise position.

TONE
Changes the tone of the sound from bass to treble.

Figure 1–2. Typical operating controls for a color-TV receiver. *(Courtesy of Heath Company)*

particular receiver section, as seen in the block diagram of Fig. 1–4. Each control varies the circuit action of its associated section in some manner. For this reason, a technician checks the responses of certain controls in a preliminary approach to sectionalization of the picture and/or sound trouble symptom(s). It is in-

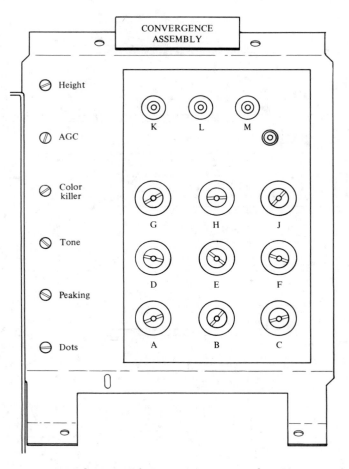

Figure 1–3. A typical group of maintenance controls. (*Courtesy of Heath Company*)

structive briefly to note the function of each shaded block depicted in Fig. 1–4. From the video amplifier, the chroma signal branches off into the chroma bandpass amplifier, where its amplitude is stepped up and the black-and-white signal is rejected. The video signal is applied to the delay line by the video amplifier, where it undergoes a time delay of approximately 1 μs. From the delay line, the video signal is applied to the video-output stage.

The chroma signal branches off into the burst amplifier. In the burst amplifier, the color burst is stepped up in amplitude, and the black-and-white signal is rejected.

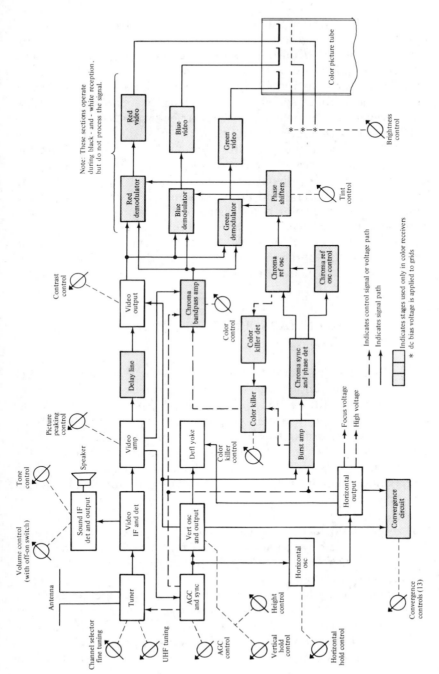

Figure 1-4. Block diagram of color-TV receiver, with operating and maintenance controls.

Next, the chroma signal from the chroma bandpass amplifier in Fig. 1–4 is applied to red, green, and blue demodulators, where it is combined with the reconstituted color subcarrier (3.58-MHz subcarrier) and mixed with the black-and-white signal. The color subcarrier is reconstituted by processing the color-burst signal through the chroma sync and phase-detector section for locking the chroma reference oscillator (subcarrier oscillator). In turn, the output from the chroma oscillator is split into three suitable phases for injection into the three chroma demodulators. From the chroma demodulators, the red video signal, the blue video signal, and the green video signal are stepped up in amplitude by the color video amplifiers, and are then applied to the color picture tube. Note that the color-killer system senses whether a color burst is present, and automatically disables the chroma bandpass amplifier if an incoming color signal is not present. The convergence circuit brings the three electron beams in the color picture tube into proper focus, as detailed subsequently.

1 . 2   NORMAL AND ABNORMAL CONTROL RESPONSES

Preliminary analysis of a trouble symptom in a color-TV receiver requires a good knowledge of normal and abnormal control responses. As an illustration, if the receiver has lost color sync, as shown in Fig. 1–1(c), it may be possible to correct the trouble by readjusting the horizontal-hold control. On the other hand, if the picture cannot be brought into sync by control adjustment, the control is said to be *out of range*. This condition is caused by some component defect in the horizontal-oscillator section, or in its associated AGC and sync section, pictured in Fig. 1–4. At this point, it is instructive to note that color sync is a subfunction of horizontal sync lock. In other words, it is possible to encounter loss of color sync without loss of horizontal sync, as seen in Fig. 1–1(b). However, if horizontal sync is lost, color sync will also be lost, as shown in Fig. 1–1(c).

When the picture has broken horizontal sync lock, it is sometimes possible to bring the picture into sync by turning the hold control to a critical setting. In such a case, the horizontal-hold control is said to lack normal locking range. This is another trouble symptom that indicates some component defect in the sync section, or in the horizontal-oscillator section. Although it is less likely, there is also a possibility that critical sync lock is being caused by technical difficulties at the transmitter. To check this possibility, the channel selector may be set to another channel. Occasionally, excessive ripple voltage on the V_{cc} (B+) lines will cause critical sync lock. Spurious AC voltages on the supply line can be quickly investigated with an oscilloscope, such as that illustrated in Fig. 1–5. Spurious AC voltages usually result from an open decoupling capacitor along the supply line, or a defective filter capacitor in the power supply.

Next, it is helpful to consider the normal responses that are observed when the fine-tuning control is adjusted. With reference to Fig. 1–6, the color picture has maximum quality when the fine-tuning control is correctly adjusted. If the fine-tuning control is turned too far left, the color "drops out" of the picture, leaving only a black-and-white image. On the other hand, if the fine-tuning control is turned too far right, both the color image and the

Figure 1–5. A high-performance oscilloscope for color-TV servicing. *(Courtesy of B&K Precision, a branch of Dynascan Corporation)*

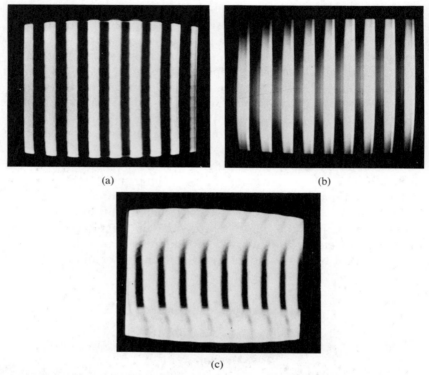

Figure 1–6. Normal effect of fine-tuning control variation: **(a)** Color picture with correct tuning; **(b)** Control turned too far left; color is lost; **(c)** Control turned too far right; both black-and-white and color are lost.

black-and-white image disappear, leaving a random sound-and-noise interference display. These are the responses that are observed in a normally operating receiver. If the color portion of the image is not displayed at any setting of the fine-tuning control, for example, a fault in the color signal channel is indicated. Again, if the best picture is obtained at one setting of the fine-tuning control, but the best sound reproduction is obtained at another setting of the fine-tuning control, another kind of malfunction is indicated in the picture channel. Analysis of these trouble symptoms is detailed subsequently.

Referring to Fig. 1–4, it is instructive to note the normal response to variation in setting of the color control. These responses are illustrated in Fig. 1–7. If normal response is not obtained from

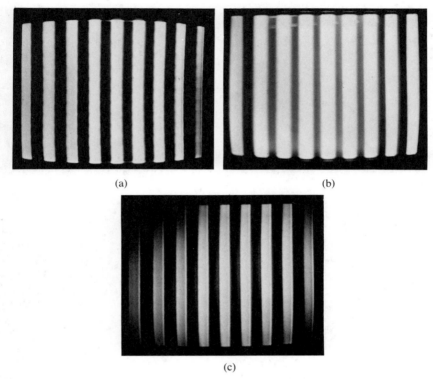

(a)   (b)

(c)

Figure 1–7. Normal responses to adjustment of the color control: **(a)** Control set correctly; **(b)** Control turned too high; **(c)** Control turned too low.

adjustment of the color control, it is indicated that there is a defect in the chroma channel. Although the color control operates in the chroma bandpass amplifier section, the fault may be elsewhere, such as in the video amplifier, or in the chroma oscillator network. Therefore, chroma signal-tracing procedures would be used to localize the difficulty. With reference to Fig. 1–8, correct color reproduction is normally obtained at a particular setting of the tint control, with distorted color reproduction on either side of the correct setting. Inability to obtain correct color reproduction by adjustment of the tint control points to a fault in the phase-shifter section, or in the color demodulator section(s). Accordingly, oscilloscope tests would be made to localize the trouble area.

Adjustment of the color-killer control normally permits color-

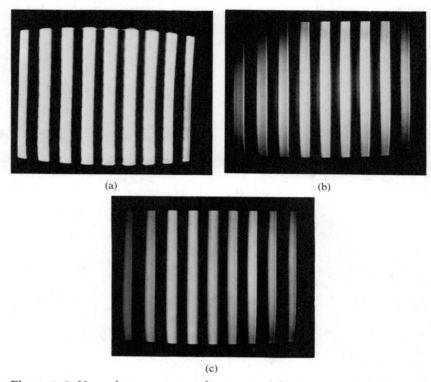

Figure 1–8. Normal responses to adjustment of the tint control: **(a)** Control set correctly; **(b)** Control turned too high; **(c)** Control turned too low.

picture reproduction up to a certain point, past which the color drops out of the picture and a black-and-white image is displayed. The color killer is adjusted to a point that almost kills color reproduction on a weak signal. This ensures that colored noise (confetti) will not interfere with the image when the receiver is tuned to an incoming black-and-white signal. When defects occur in the color-killer section, it may be impossible to reproduce a color image at any setting of the control, or it may be impossible to kill the chroma component in a color picture. In most cases, the fault will be tracked down to a defective component, such as an "open" or "shorted" capacitor in the color-killer section.

With reference to Fig. 1–4, the convergence controls are adjusted to bring all three electron beams to focus at the same point

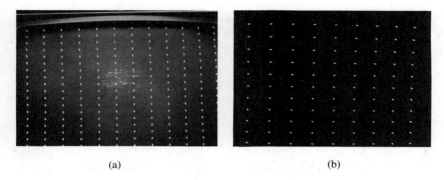

(a)   (b)

Figure 1–9. Effect of convergence-control adjustments: **(a)** Electron beams correctly converged; **(b)** Example of misconvergence.

on the picture-tube screen. The effect of misconvergence is illustrated in Fig. 1–9. Convergence procedure requires practical experience as well as study, and is the most involved maintenance procedure in color-receiver service. Note that there is a trend toward the use of color picture tubes that are permanently converged at the time of manufacture. These picture tubes are comparatively expensive, because they are supplied with permanently mounted convergence and deflection coils on the picture-tube neck. Inasmuch as the majority of color-TV receivers in current use are provided with convergence controls, it is essential that the technician have a good knowledge of convergence procedure. This topic is detailed in a later chapter.

1 . 3   SNOW, CONFETTI, AND AUDIO NOISE LEVELS

Video snow, confetti, and audio noise levels may provide useful data for preliminary trouble localization when the symptoms are weak or no picture, weak or no chroma image, and/or weak or no sound. Figure 1–10 illustrates three different snow levels and a typical confetti display in the raster. Analysis of the snow level is based on the fact that the earlier stages in the signal channel contribute more snow than the later stages. In other words, the snow (noise) output from the tuner (Fig. 1–4) is amplified through the IF section and through the video-amplifier section. Therefore, a fault that stops the signal in the tuner section causes a much higher snow level than a fault that stops the signal in the video-

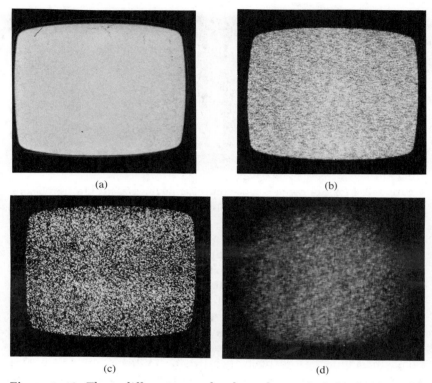

(a)   (b)

(c)   (d)

Figure 1–10. Three different snow levels, and a confetti display: **(a)** Low snow level; **(b)** Medium snow level; **(c)** High snow level; **(d)** Confetti display.

detector section. Snow is most prominent on a vacant channel, in normal operation, because the AGC section is then "wide open." As would be anticipated, snow is also most prominent when the contrast control is advanced to maximum. Note that in case of AGC trouble, a picture channel that is in good operating condition may not produce any snow display. Therefore, in case of doubt, the AGC voltage should be measured, and the AGC line clamped, if necessary. As a rough rule of thumb, the picture channel is capable of reproducing a good picture if the snow level is normal and the "snowflakes" appear to be sharply focused.

Confetti is normally displayed prominently on a vacant channel when the color-killer control is turned to the end of its range to enable the chroma channel. Note that a confetti pattern normally appears somewhat coarser than a snow pattern, because the bandwidth of the chroma channel is considerably less than the bandwidth of the video-amplifier channel. If no confetti pattern can be displayed, it may be concluded that the receiver is incapable of displaying a color image. Audio noise levels are determined by the same general principles as snow levels. In other words, a defect that stops the signal in the tuner section causes a much higher audio noise level than a defect that stops the signal in the intercarrier-IF section (Fig. 1–4). In normal operation, audio noise is most evident on a vacant channel, with the volume control turned to maximum. All preliminary trouble-localization procedures must be cross-checked and supplemented, as explained subsequently.

1.4 PROCEDURAL TROUBLESHOOTING APPROACH

Considerable time and effort can usually be saved if a logical troubleshooting procedure is followed. An effective general procedure for troubleshooting a color-TV receiver is depicted in Fig. 1–11. First, the onset and nature of the trouble symptoms should be discussed with the set owner. In some cases, the information that is obtained discloses the nature of the defect, and the necessity for making systematic tests is avoided. As an illustration, if the set owner states that the receiver started operating poorly after he attempted to improve the alignment of the tuned circuits, it is evident that a realignment of the receiver is probably all that will be required to restore normal operation. On the other hand, if the set owner cannot provide definite clues concerning trouble symptoms, the operation of the receiver should be observed carefully. A thorough visual inspection can also provide helpful clues. As an example, suppose that the picture wavers and bends and appears generally distorted, with "touchy" horizontal sync lock. In such a case, an experienced technician would suspect that the power supply is defective, and that replacement filter capacitors are probably required.

Occasionally, a component defect can be located by noting an overheated component, sometimes accompanied by a burning odor. For example, if a composition resistor overheats owing to excessive current flow, it is likely to produce a distinctive burning odor. The resistor will also appear dark and charred after substantial overheating. A component defect can occasionally be lo-

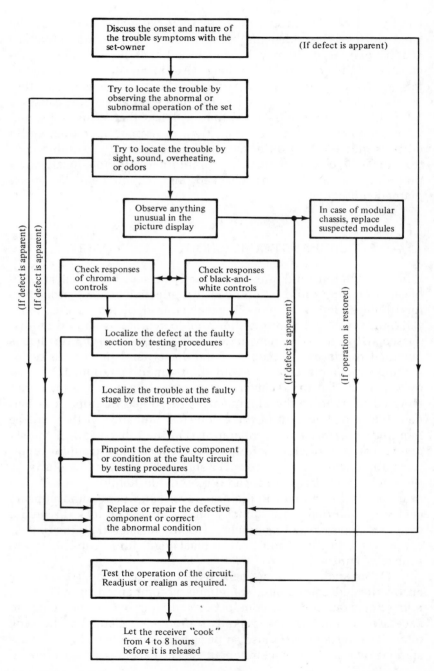

Figure 1–11. General procedure for troubleshooting a color-TV receiver.

cated by sound. As an illustration, excessive current demand from a power transformer results in a prominent 60-Hz hum. Arcs in the high-voltage section produce crackling and snapping sounds. When a component is replaced, it is often essential to determine why it failed. Thus, if a resistor is burned, the cause of excessive current demand must be sought. In other words, it is not likely that the receiver can be restored to normal operation by merely replacing the damaged resistor. Similarly, if a power transformer has burned out, the reason for excessive current demand must be determined before a replacement transformer is put into service.

Useful trouble clues can often be obtained by observing the action of the operating and maintenance controls. It may be found that the fault is in the black-and-white section, in the chroma section, or in the sound section. The picture display should be carefully observed for unusual features, and the sound reproduction should be noted. Fig. 1–12 shows the principal sections associated with black-and-white and sound reproduction, and Fig. 1–13 presents a general picture-raster-sound trouble symptom chart. Next, Fig. 1–14 shows the principal sections associated with color reproduction, and Fig. 1–15 presents a general chroma-section troubleshooting chart. Note that this chart is not exclusive, inasmuch as the chroma signal also passes through the tuner, video-IF amplifier, detector, and video amplifier (Fig. 1–4). Inadequate frequency response in these receiver sections can cause a "color absent" trouble symptom, although the chroma sections are operating normally. Troubleshooting procedures are detailed in following chapters.

After the defective section has been localized, instrument tests are made to identify the faulty circuit. Supplementary tests are then made to pinpoint the defective device or component. Occasionally, it may not be possible to complete this general procedure. For example, service data may be incomplete, a necessary type of test instrument may be lacking, or a highly interactive circuit may yield confusing test results. In such a case, the technician will usually resort to substitution tests. As an illustration, if it is suspected that a certain integrated circuit may be defective, and test data are inconclusive, a known good integrated circuit (IC) is substituted for the suspect. Then, if the receiver resumes normal operation, the suspicion is confirmed.

Following replacement of the defective component, any necessary "touch-up" adjustments or tuned-circuit alignment must be made. After the receiver appears to be operating normally, it is good practice to check it for marginal defects by "cooking" the

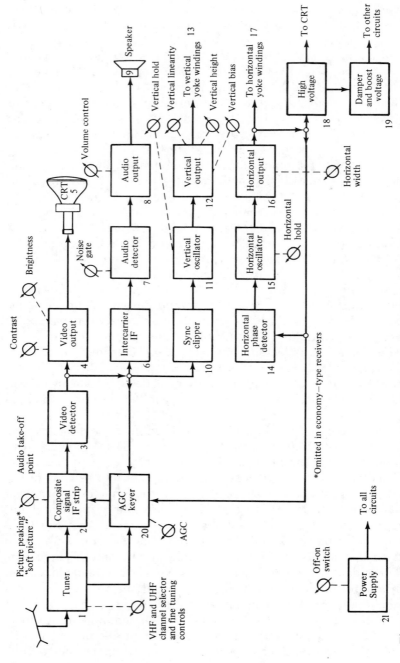

Figure 1-12. Principal sections associated with black-and-white and sound reproduction.

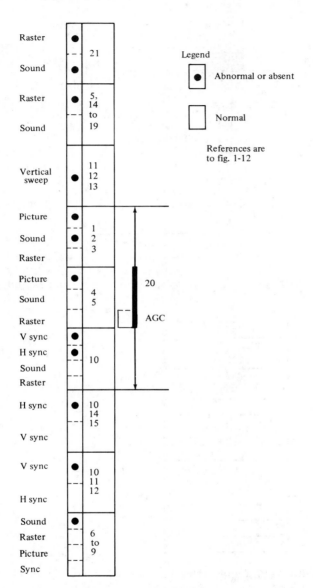

Figure 1–13. General picture-raster-sound trouble symptom chart.

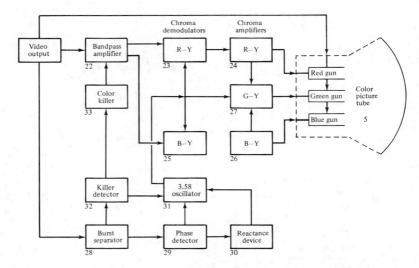

Figure 1–14. Principal sections associated with color reproduction.

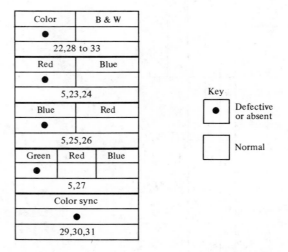

Figure 1–15. General chroma-section troubleshooting chart.

chassis for several hours. In other words, the receiver is turned on and operated continuously from four to eight hours. This precaution serves to minimize the possibility of a callback. A callback not only wastes man-hours, but also damages the reputation of a

TV service organization. Therefore, every reasonable effort is made to ensure that a repaired receiver will "stay repaired" as long as possible..

1.5 BASIC COLOR-TV TEST EQUIPMENT

A volt-ohmmeter is the most basic color-TV test instrument. A hi-lo field-effect multimeter, such as that pictured in Fig. 1–16, is de-

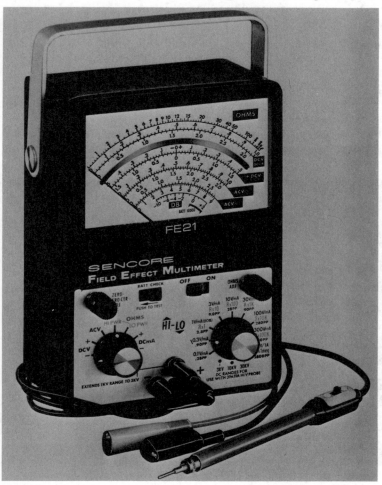

Figure 1–16. A hi-lo field-effect multimeter *(Courtesy of Sencore, Inc.)*

sirable for checking solid-state circuitry. It has an input resistance of 15 megohms, and a full-scale value of 0.1 volt on its first DC-voltage range. Two resistance-measuring functions are provided. One, the "hi" function, is conventional. The other, the "lo" function, applies less than 0.1 volt across the terminals under test. This feature enables resistance values to be measured in circuit, inasmuch as the test voltage is too low to turn on the junction of a normal semiconductor diode or transistor. This topic is detailed subsequently.

An oscilloscope such as that illustrated in Fig. 1–17 is also a basic color-TV service instrument. Wide-band vertical-amplifier response is essential, so that the 3.58-MHz color burst and related chroma waveforms can be displayed at true peak-to-peak voltages. Thus, the oscilloscope in this example is rated for a vertical-amplifier frequency response that is less than −3 dB (30 percent down, approximately) at 10 MHz. Triggered sweep is provided; although this feature is not essential, it is preferred by many technicians as a practical operating convenience. Vectorscope input and control facilities are also provided. A calibrated

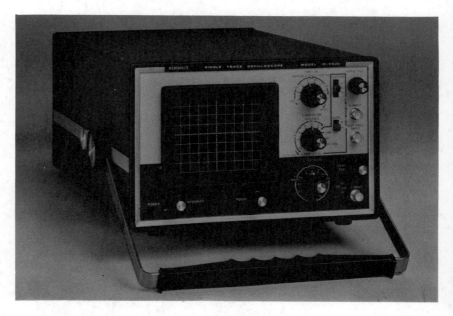

Figure 1–17. Typical oscilloscope for color-TV service. (*Courtesy of Heath Company*)

vertical step attenuator is included, to facilitate peak-to-peak voltage measurements of waveforms.

A keyed-rainbow color-bar and pattern generator, such as that pictured in Fig. 1–18, is another basic instrument for color-TV service. This type of generator supplies a color-bar signal for troubleshooting chroma circuitry, plus vertical-bar, horizontal-bar, crosshatch, and dot patterns that are used in picture-tube convergence procedures. Typical white-dot, crosshatch, and chroma-bar patterns are shown in Fig. 1–19. A suitable generator is also

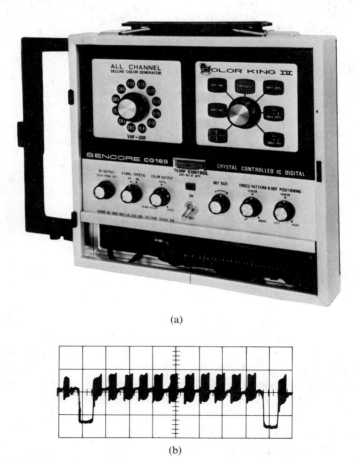

(a)

(b)

Figure 1–18. Keyed-rainbow color-bar and pattern generator: **(a)** Appearance of generator; **(b)** Color-bar video waveform output. *(Courtesy of Sencore, Inc.)*

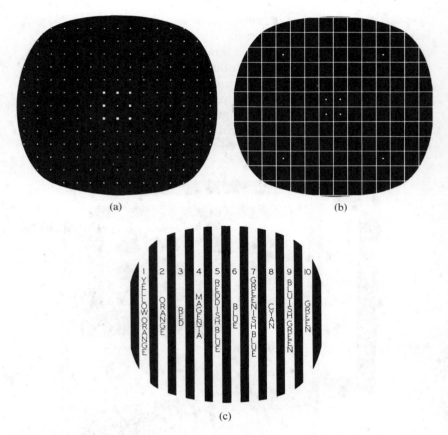

Figure 1–19. Typical screen patterns: **(a)** white-dot; **(b)** crosshatch; **(c)** chroma bar.

needed for tuned-circuit alignment procedures, such as that illustrated in Fig. 1–20. This is an example of a sweep-frequency generator with a built-in marker generator. It is used in combination with an oscilloscope to display RF, IF, and VF (video-frequency or chroma) response curves.

In addition to the foregoing instruments, color-TV technicians employ numerous other types of test equipment, such as capacitor checkers, semiconductor testers or curve tracers, high-voltage probes (Fig. 1–21), tube testers, signal-substitution generators of various designs, RC substitution boxes, and bench power supplies. Some technicians also utilize color picture-tube test jigs such as that in Fig. 1–22. The jig contains a standard color picture

Figure 1–20. A sweep-frequency generator with a built-in marker generator. (*Courtesy of B&K Precision Mfg. Co. Division of Dynascan Corporation*)

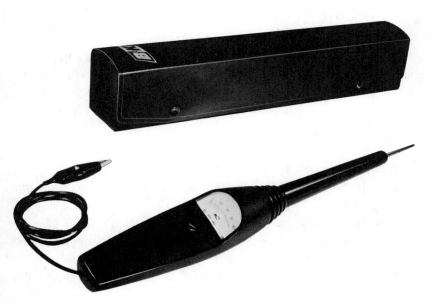

Figure 1–21. A high-voltage probe with built-in meter. (*Courtesy of B&K Precision Mfg. Co. Division of Dynascan Corporation*)

Figure 1–22. A color picture-tube test jig. *(Courtesy of RCA)*

tube that is connected with cables to the color chassis being serviced. It provides an advantage in that it shows quickly whether
the trouble is in the receiver circuitry or in the color picture tube.

REVIEW QUESTIONS

1. What is the basic troubleshooting approach?
2. Can the same trouble symptom be caused by more than one
   kind of fault?
3. How is the color subcarrier reconstituted?
4. Is it possible for color sync to be lost without concurrent loss
   of horizontal sync?
5. Could horizontal sync be lost without concurrent loss of color
   sync?
6. Why is the color portion of an image sometimes rejected by
   the front end?

7. Define the term *confetti*.

8. Explain the meaning of *convergence*.

9. What is the first step in a logical troubleshooting procedure?

10. Describe the chief features of a hi-lo multimeter.

11. Name the chief requirements for an oscilloscope used in color-TV troubleshooting procedures.

12. State the signal outputs typically provided by a keyed-rainbow and pattern generator.

13. What is the chief application for a sweep-frequency and marker generator?

14. Are tube testers employed in present-day color-TV service shops?

15. Explain the function of a color picture-tube test jig.

Color-reproduction Troubles in the Black-and-white Sections

2.1 GENERAL CONSIDERATIONS

A color-TV receiver includes all of the conventional black-and-white circuitry with which we are familiar, plus a chroma section with a color picture tube, as shown in Fig. 2–1. In case the RF section, IF section, or video section develops poor high-frequency response, owing to a component defect, the chroma signal may be weakened or suppressed. Accordingly, the picture is displayed with little or no color, although the black-and-white image may still be acceptable. Or, if horizontal sync lock is marginal, color-sync action usually becomes unstable. Some kinds of black-and-white circuit malfunctions cause poor color "fit," although the effect does not distort the black-and-white image. Therefore, it is essential for the color-TV technician to understand the details of color signal processing in the black-and-white sections of a color-TV receiver.

Frequency interrelations of the black-and-white picture-channel sections are shown in Fig. 2–2. Since the color-subcarrier frequency is 3.58 MHz, it is evident that excessive signal attenuation at this frequency, whether in the RF tuner, the video-IF section, or the video-amplifier section, will seriously weaken both the chroma signal and the color burst. A substantially weakened chroma signal produces pale and washed-out colors in the image;

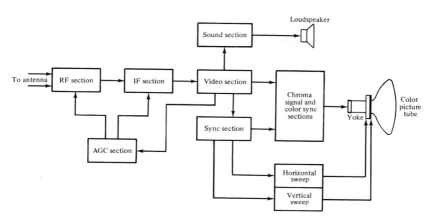

Figure 2-1. Chroma signal can be affected by malfunctions of the black-and-white sections.

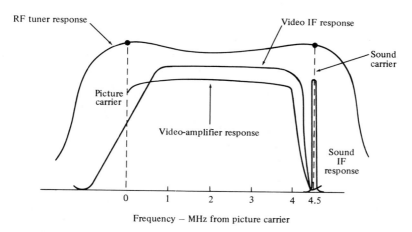

Figure 2-2. Frequency interrelations of the black-and-white picture-channel sections.

a seriously attenuated color burst results in unstable color sync lock, or complete loss of color sync. Table 2–1 lists the VHF and UHF channel frequencies, and Fig. 2–3 shows the picture-channel frequency interrelations for channel 4 and channel 10 reception.

In the example of Fig. 2–3, the picture carrier falls at 45.75 MHz, and the color subcarrier falls at 42.17 MHz. Thus, the

TABLE 2–1

FCC Television Channel Allocations

P=Picture Carrier Freq.(MHz)
S=Sound Carrier Freq.(MHz)
FREQ. LIMITS OF CHANNEL (MHz)

Channel	P Freq. (MHz)	S Freq. (MHz)	Freq. Limits (MHz)
2	55.25	59.75	54 – 60
3	61.26	65.76	60 – 66
4	67.25	71.75	66 – 72
5	77.25	81.75	76 – 82
6	83.25	87.75	82 – 88
7	175.25	179.75	174 – 180
8	181.25	185.75	180 – 186
9	187.25	191.75	186 – 192
10	193.25	197.75	192 – 198
11	199.25	203.75	198 – 204
12	205.25	209.75	204 – 210
13	211.25	215.75	210 – 216
14	471.25	475.75	470 – 476
15	477.25	481.75	476 – 482
16	483.25	487.75	482 – 488
17	489.25	493.75	488 – 494
18	495.25	499.75	494 – 500
19	501.25	505.75	500 – 506
20	507.25	511.75	506 – 512
21	513.25	517.75	512 – 518
22	519.25	523.75	518 – 524
23	525.25	529.75	524 – 530
24	531.25	535.75	530 – 536
25	537.25	541.75	536 – 542
26	543.25	547.75	542 – 548
27	549.25	553.75	548 – 554
28	555.25	559.75	554 – 560
29	561.25	565.75	560 – 566
30	567.25	571.75	566 – 572
31	573.25	577.75	572 – 578
32	579.25	583.75	578 – 584
33	585.25	589.75	584 – 590
34	591.25	595.75	590 – 596
35	597.25	601.75	596 – 602
36	503.25	607.75	602 – 608
37	609.25	613.75	608 – 614
38	615.25	619.75	614 – 620
39	621.25	625.75	620 – 626
40	627.25	631.75	626 – 632
41	633.25	637.75	632 – 638
42	639.25	643.75	638 – 644
43	645.25	649.75	644 – 650
44	651.25	655.75	650 – 656
45	657.25	661.75	656 – 662
46	663.25	667.75	662 – 668
47	669.25	673.75	668 – 674
48	675.25	679.75	674 – 680
49	681.25	685.75	680 – 686
50	687.25	691.75	686 – 692
51	693.25	697.75	692 – 698
52	699.25	703.75	698 – 704
53	705.25	709.75	704 – 710
54	711.25	715.75	710 – 716
55	717.25	721.75	716 – 722
56	723.25	727.75	722 – 728
57	729.25	733.75	728 – 734
58	735.25	739.75	734 – 740
59	741.25	745.75	740 – 746
60	747.25	751.75	746 – 752
61	753.25	757.75	752 – 758
62	759.25	763.75	758 – 764
63	765.25	769.75	764 – 770
64	771.25	775.75	770 – 776
65	777.25	781.75	776 – 782
66	783.25	787.75	782 – 788
67	789.25	793.75	788 – 794
68	795.25	799.75	794 – 800
69	801.25	805.75	800 – 806
70	807.25	811.75	806 – 812
71	813.25	817.75	812 – 818
72	819.25	823.75	818 – 824
73	825.25	829.75	824 – 830
74	831.25	835.75	830 – 836
75	837.25	841.75	836 – 842
76	843.25	847.75	842 – 848
77	849.25	853.75	848 – 854
78	855.25	859.75	854 – 860
79	861.25	865.75	860 – 866
80	867.25	871.75	866 – 872
81	873.25	877.75	872 – 878
82	879.25	883.75	878 – 884
83	885.25	889.75	884 – 890

Redrawn by permission of Howard W. Sams & Co., Inc.

chroma signal is processed at the same amplitude as the picture carrier through the IF section. Next, it is instructive to consider how the color burst becomes attenuated if the IF amplifier has subnormal high-frequency response. As shown in Fig. 2–4, the

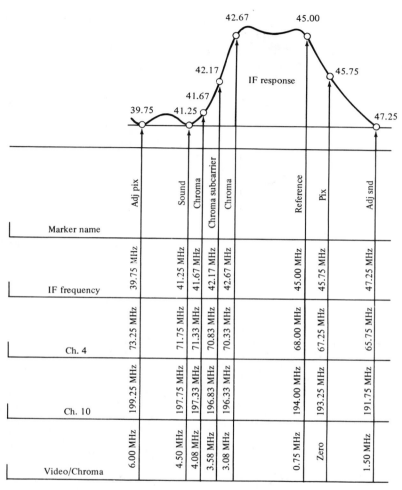

Figure 2–3. A widely used contour for the IF response curve, depicting frequency interrelations. (*Redrawn by permission of B & K Mfg. Co., Division of Dynascan Corporation*)

sync-pulse and color-burst waveforms are processed on opposite sides of the IF response curve. The color burst has a frequency of 3.58 MHz. On the other hand, the horizontal sync pulse comprises low frequencies flanking the picture carrier. The highest significant frequency in the sync pulse is approximately ten times its fundamental frequency, or .158 MHz. Therefore, when the higher IF frequencies are attenuated, as exemplified in Fig. 2–5, the color burst is attenuated accordingly, although the horizontal sync-pulse waveform is virtually unaffected.

These considerations also apply to the RF section and to the video-amplifier section. In other words, if the RF tuner has poor

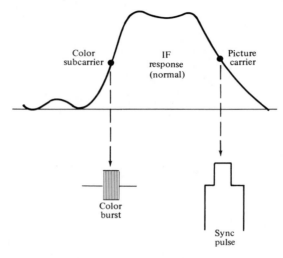

Figure 2–4. Sync-pulse and color-burst waveforms are processed on opposite sides of the IF response curve.

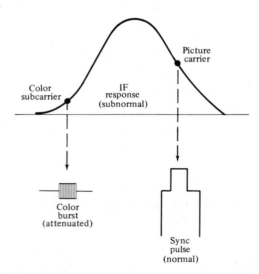

Figure 2–5. Narrow IF bandwidth causes attenuation of the color burst.

high-frequency response, the color burst (and chroma signal) will be attenuated in the same manner as shown in Fig. 2–5. Similarly, if the video-amplifier section has poor high-frequency response, the 3.58-MHz signal components will be attenuated accordingly. Poor high-frequency response in the RF tuner or IF section results from misalignment. In other words, component defects such as open capacitors can cause frequency shifts in tuned circuits, with resulting misalignment. Poor high-frequency response in the video-amplifier section can be caused by off-value collector load resistors or by defective peaking coils. Occasionally, a peaking coil is encountered that is not defective, but its inductance is incorrect, owing to a previous replacement error.

Consider a trouble symptom of no picture with sound normal. Preliminary localization can be made by progressive signal-injection tests, as tabulated in Fig. 2–6. The tests noted are referenced to the skeleton circuit shown in Fig. 2–7. Next, consider a

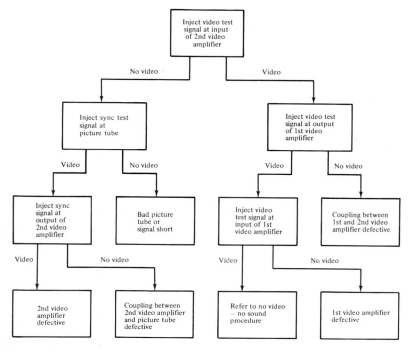

Figure 2–6. Progressive signal-injection tests for a trouble symptom of no picture with sound normal. *(Adapted from an original by B & K Mfg. Co. Division of Dynascan Corporation)*

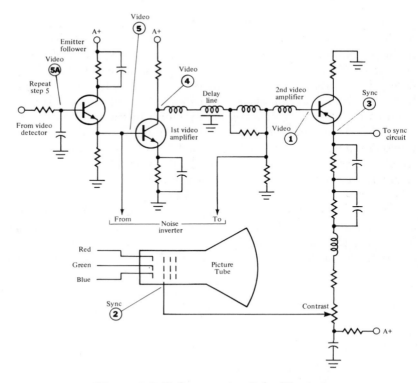

Figure 2–7. Reference circuit for Fig. 2–6.

trouble symptom of no picture and no sound. Signal-injection tests are made as summarized in Fig. 2–8. These tests are referenced to the skeleton circuit depicted in Fig. 2–9. Again, consider a trouble symptom of no sound with normal picture. Progressive signal-injection tests are made as listed in Fig. 2–10. These tests are referenced to the skeleton circuit shown in Fig. 2–11.

2.2   COLOR-SYNC MALFUNCTION CAUSED BY MISTIMED GATING PULSE

As shown in Fig. 2–12, the color burst is removed from the horizontal-sync pulse by a gating pulse. This gating pulse is derived in the same manner as the AGC gating pulse from the horizontal-deflection flyback pulse. When the horizontal-hold con-

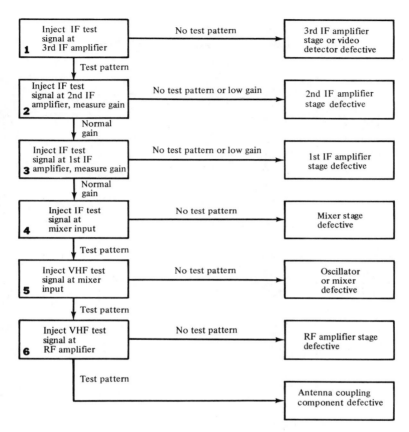

Figure 2–8. Progressive signal-injection tests for a trouble symptom of no picture and no sound. (*Adapted from an original by B & K Mfg. Co. Division of Dynascan Corporation*)

trol is adjusted in many receivers, the picture image tends to shift somewhat to the left or right on the picture-tube screen. In turn, the flyback pulse is pulled more or less out of time with the horizontal-sync pulse, and the burst-gating pulse becomes mistimed accordingly. If the mistiming is marginal, a portion of the color burst is lost. In turn, color-sync lock becomes unstable, and the colors in the displayed picture vary erratically. Again, if the mistiming is severe, color sync lock is broken and the colors in the displayed picture form a series of diagonal rainbows.

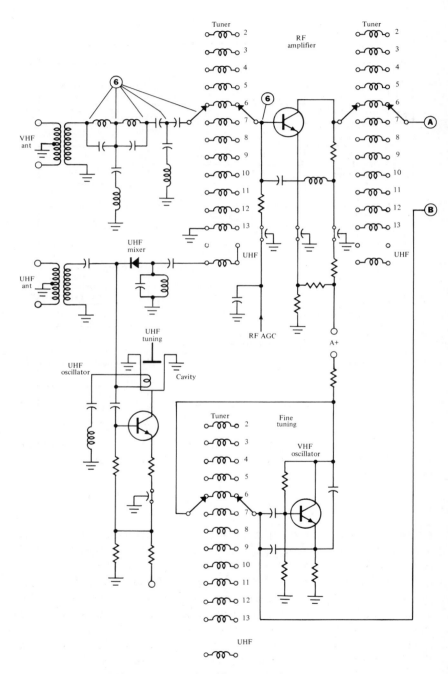

Figure 2–9. Reference circuit for Fig. 2–8.

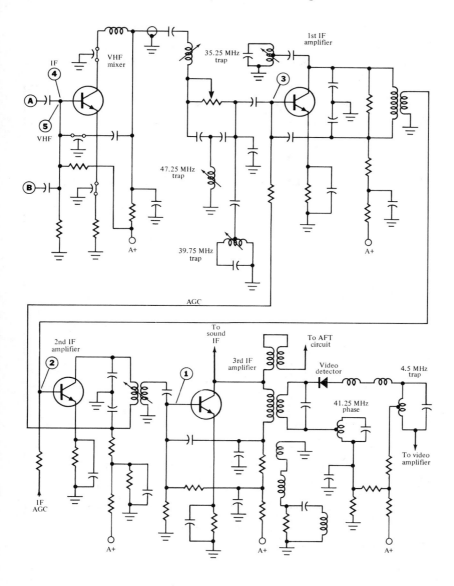

Figure 2–9. Continued.

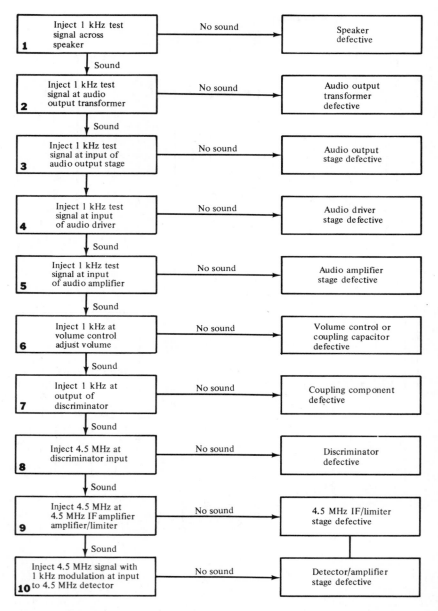

Figure 2–10. Progressive signal-injection tests for a trouble symptom of no sound with normal picture. *(Adapted from an original by B & K Mfg. Co. Division of Dynascan Corporation)*

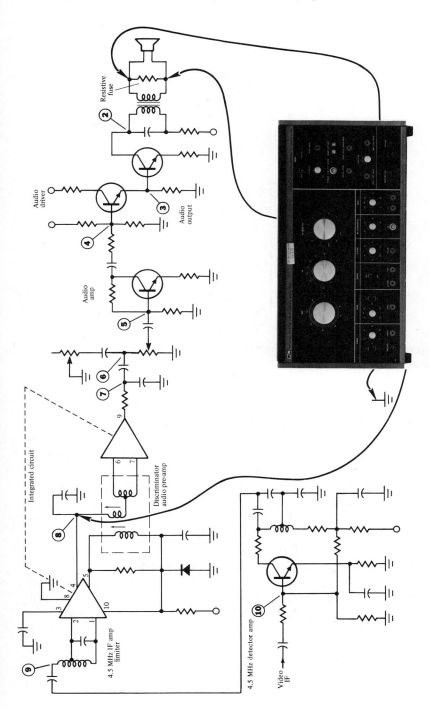

Figure 2–11. Reference circuit for Figure 2–10.
(Adapted from an original by B & K Mfg. Co. Division of Dynascan Corporation)

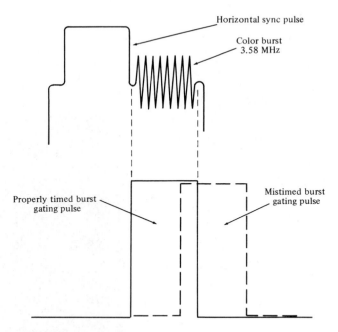

Figure 2–12. Color burst is removed from the horizontal-sync pulse by means of a gating pulse.

Sometimes a component defect, such as a leaky capacitor, in the horizontal-sync section will shift the range of the horizontal-hold control, so that the picture can be locked horizontally only when the hold control is turned to an extreme end of its range. In such a case, color-sync action is likely to be affected. Therefore, the technician should make certain that the operation of the horizontal-sync section is normal, before proceeding to check out the color-sync section. It is also advisable to check the waveform of the burst gating pulse with an oscilloscope. As shown in Fig. 2–13, it may be found that the burst-gating pulse is narrow and sharply peaked, instead of having normal width and a reasonably flat top. This fault results in passage of a fraction of the color burst, instead of the entire burst. In case of gating-pulse waveform distortion, the components in the burst-gating network should be checked for correct resistance, capacitance, and inductance values.

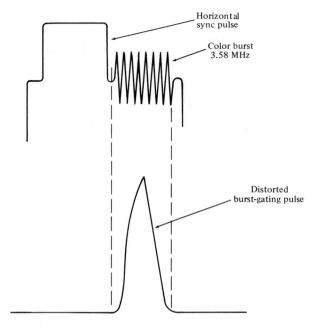

Figure 2–13. A narrow and sharply peaked gating pulse impairs color-sync action.

2.3 ABNORMAL RESPONSE IN THE 3.1–4.1 MHz INTERVAL

Instead of attenuated high-frequency response, an IF amplifier sometimes develops abnormal response over a small interval owing to regeneration. If a sharp high peak occurs in the chroma-signal region of the IF curve, serious color distortion can occur. The chroma signal occupies a frequency interval from 3.1 to 4.1 MHz apart from the picture carrier. With reference to Fig. 2–14, the chroma signal extends from 41.65 MHz to 42.75 MHz. If the IF section becomes regenerative, and a sharp high peak is developed in this interval, as exemplified by the dotted lines, serious hue shift occurs. In other words, this condition causes a large change in chroma-signal phase over the peaked excursion. Since chroma-

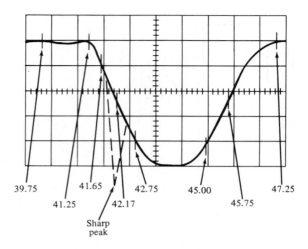

39.75   41.65   42.75   45.00   47.25
    41.25   42.17   45.75

Sharp
peak

Figure 2–14. Typical response-curve distortion caused by regeneration.

signal phase corresponds to hue, the result is serious changes in the reproduced colors.

Intermediate frequency regeneration is often caused by open bypass or decoupling capacitors. An open decoupling capacitor permits feedback to take place within the IF network. Again, an open bypass capacitor will sometimes change the characteristics of an IF stage substantially, with resulting frequency distortion. Open bypass capacitors along an AGC line are very likely to cause regeneration. In some receivers, IF misalignment can also cause regeneration. In other words, if the input and output tuned circuits of an individual stage are peaked to the same frequency, it may happen that regeneration results. Therefore, IF alignment in a color-TV receiver should be carried out in strict accordance with the specifications in the receiver service data. Note that regeneration can also occur in RF tuners and in video amplifiers. However, the condition is generally more troublesome in the IF section, because of its comparatively high gain.

Figure 2–15 shows the basic relations between hue and phase. Each hue has the same frequency (3.58 MHz), but it has a distinctive phase with respect to the color burst. Thus, yellow has a chroma phase angle of approximately 10 degrees, red has a 76-degree phase angle, magenta has a 120-degree phase angle, blue has about a 190-degree phase angle, cyan has a 256-degree phase

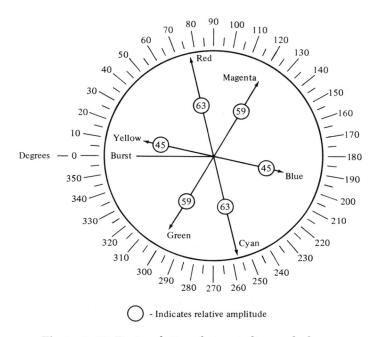

○ - Indicates relative amplitude

Figure 2–15. Basic relations between hue and phase.

angle, and green has a 300-degree phase angle. These relative phase angles can be preserved in processing through the picture-signal channel only if the RF, IF, and video-amplifier sections have reasonably normal frequency-response curves. It is instructive to note that phase is related to frequency response in the general manner seen in Fig. 2–16. As the frequency varies from one end of the passband to the other, the phase angle of the current varies from 90 degree lagging to 90 degree leading. Phase relations are usually checked with a vectorscope, as detailed subsequently. Figure 2–17 shows an example of a vectorgram.

Good color reproduction requires that the tuned circuits in a color-TV receiver operate properly in combination. Thus, a tilted frequency response and a nonlinear phase characteristic in one receiver section can be compensated by an oppositely tilted frequency response with a suitable phase characteristic in another receiver section. With reference to the frequency-response curves shown in Fig. 2–18, the tilt in the IF response curve through the chroma-signal interval is compensated by an opposite tilt in the

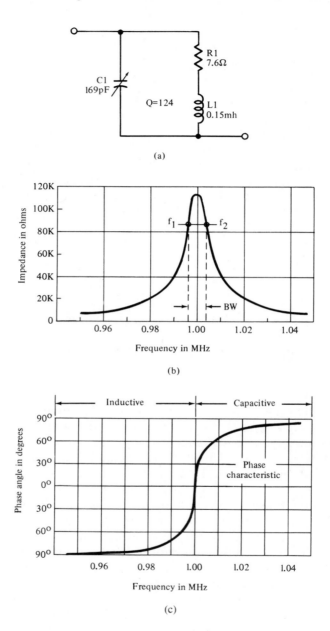

(a)

(b)

(c)

Figure 2–16. Frequency-response curve and phase characteristic for a simple parallel-resonant circuit: **(a)** Configuration; **(b)** Frequency response; **(c)** Phase characteristic.

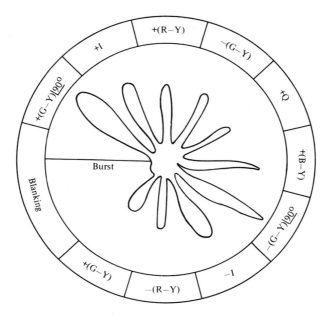

Figure 2–17. Typical demodulation phase and demodulated phases displayed by a keyed-rainbow vectorgram. (*Courtesy of Sencore, Inc.*)

bandpass-amplifier response curve. In turn, the overall IF and bandpass-amplifier response curve is reasonably flat-topped and also has an acceptably linear phase characteristic. The overall frequency response of a color-TV receiver is checked to best advantage by means of the video-sweep modulation (VSM) method. As explained in detail in a later chapter, this alignment method employs a specialized sweep-frequency signal that is applied at the input of the RF tuner, with an oscilloscope connected at the output of the bandpass amplifier.

2.4 SIGNAL COMPRESSION OR CLIPPING

Signal compression or clipping can occur in the black-and-white section of a receiver, particularly in stages that operate at a fairly high level, such as the video amplifier. Since the video amplifier processes the complete color signal (black-and-white signal plus the chroma signal), an overload condition in this stage affects both

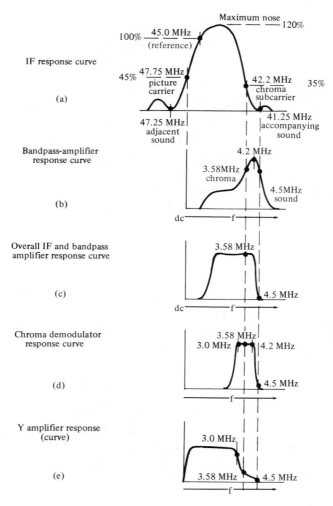

Figure 2–18. Frequency response curves for various sections of a color-TV receiver. (*Courtesy of General Electric Company*)

color and black-and-white picture reproduction. Overload can be caused by AGC malfunction resulting in excessive output from the picture detector. Again, overload can occur at the normal signal level owing to incorrect bias voltage on a transistor. The first video-amplifier transistor usually operates in the common-collector (emitter-follower) mode. In case the bias voltage is incor-

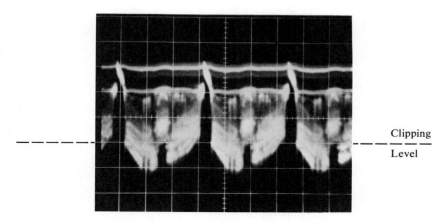

Clipping
Level

Figure 2–19. Video signal highlights are lost and blurred when the waveform is clipped at the indicated level.

rect and the signal peaks drive the transistor past cutoff, the signal waveform will be sharply clipped in proportion.

It is instructive to note the effect of clipping in a black-and-white video waveform within the white and light-gray region. With reference to Fig. 2–19, clipping at the indicated level results in loss of highlights and blurring of the bright detail in the picture image. Thus, the clipping action causes the picture to appear dim, with highlights that look "muddy" and "filled up." Next, consider the effect of clipping in a composite color-video signal. The composite signal has both a black-and-white component and a chroma component. Clipping affects the black-and-white component as noted above, and also causes color distortion. In Fig. 2–20, it will be observed that clipping of the composite color-bar signal at the indicated level will result in partial loss of yellow and cyan chroma signals. In turn, these colors will appear weak and "washed out" in the picture image.

2.5 SUPPLEMENTARY COLOR-TV TEST EQUIPMENT

Many technicians use a television analyzer such as that illustrated in Fig. 2–21 for signal-injection tests in color-TV receivers. A television analyzer is an elaborate and specialized type of signal generator that provides a test-pattern signal, a keyed-rainbow signal, a

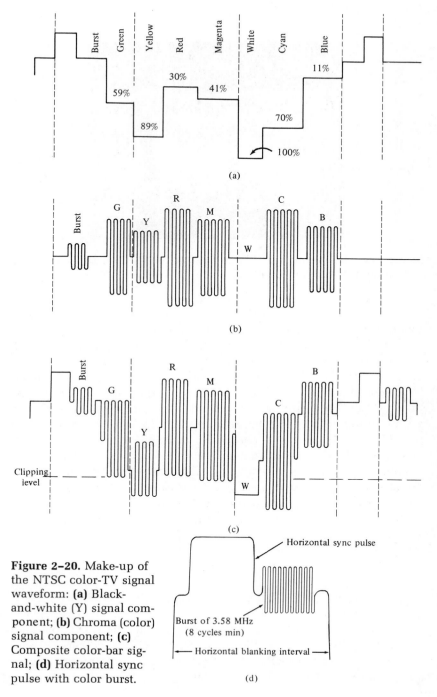

Figure 2–20. Make-up of the NTSC color-TV signal waveform: **(a)** Black-and-white (Y) signal component; **(b)** Chroma (color) signal component; **(c)** Composite color-bar signal; **(d)** Horizontal sync pulse with color burst.

46

Figure 2–21. A TV analyzer with facilities for chroma-circuit testing. *(Courtesy of B & K Mfg. Co. Division of Dynascan Corporation)*

white-dot signal, or a crosshatch signal at VHF, IF, video, and chroma frequencies. For example, a test signal can be injected at any point in the black-and-white picture channel, and a test pattern as shown in Fig. 2–22 will be displayed on the picture-tube screen unless there is a defective component in the path of signal flow. Similarly, a keyed-rainbow test signal can be injected at any point in the RF, IF, or video sections of the receiver, and a chroma-bar pattern as shown in Fig. 2–23 will be displayed on the picture-tube screen unless there is a defective component or device in the chroma section.

A television analyzer also provides an audio tone signal, a 4.5-MHz intercarrier sound signal that is tone-modulated, an IF sound signal, and a VHF sound signal. It supplies composite synchronizing signals for injection into each stage of the sync section. Also, vertical- and horizontal-drive signals are provided, with an AGC keying pulse, and AGC clamp-voltage source, supplemented by a boost-voltage source. The scanner section of a television analyzer provides white-dot and crosshatch patterns, as shown in Fig. 2–24, for use in convergence procedures. These flying-spot patterns are essentially the same as standard dot and crosshatch patterns. Since the flying-spot patterns are generated by scanning transparencies, any test pattern that may be desired can be obtained by making up a corresponding transparency for insertion into the analyzer.

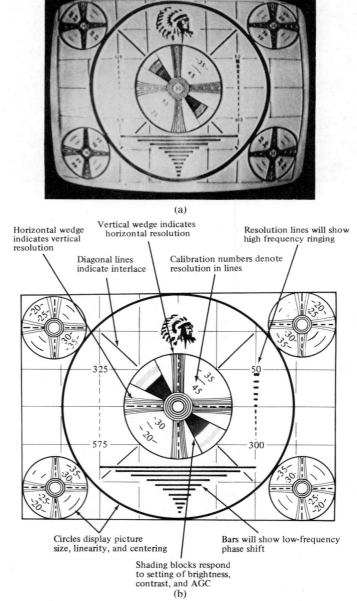

(a)

Horizontal wedge indicates vertical resolution

Vertical wedge indicates horizontal resolution

Resolution lines will show high frequency ringing

Diagonal lines indicate interlace

Calibration numbers denote resolution in lines

Circles display picture size, linearity, and centering

Bars will show low-frequency phase shift

Shading blocks respond to setting of brightness, contrast, and AGC

(b)

Figure 2–22. Test-pattern characteristics: **(a)** Normal test-pattern display on picture-tube screen; **(b)** Basic information provided by a test pattern.

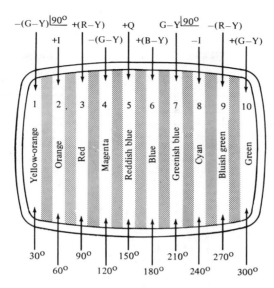

Figure 2–23. A standard chroma-bar pattern.

Smaller color-TV analyzer instruments are also available, as exemplified in Fig. 2–25. This unit provides RF, IF, sync, and sound signal outputs for signal-injection tests. Nine patterns are provided; a reference black-level pattern with sync pulses is used for color picture-tube purity checks. A single-dot pattern is provided for locating the precise center-screen point when static convergence is started; a 7 × 11 white-dot pattern (7-dot columns and 11-dot rows) is available for checking overall convergence; a single vertical-line display is employed for checking dynamic convergence at the top and bottom of the screen; a single horizontal-line display is utilized for checking dynamic convergence on the left and right sides of the screen at vertical center; a crosshair pattern is used to check deflection centering and as an aid while "roughing in" static convergence adjustments; the 7 × 11 crosshatch display is utilized for dynamic convergence checks, and in linearity, size, and overscan adjustments; the three color-bar R-Y, B-Y, -(R-Y) pattern is employed in color alignment procedures; and the gated rainbow (keyed-rainbow) pattern is used in operating tests and adjustments of the chroma section.

Color-TV laboratories use NTSC color-bar generators, as do some large service organizations. The basic difference between an

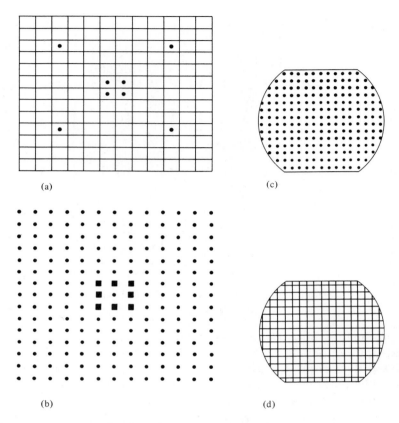

Figure 2–24. Typical white-dot and crosshatch patterns: **(a)** Flying-spot crosshatch pattern; **(b)** Flying-spot dot pattern; **(c)** Standard dot pattern; **(d)** Standard crosshatch pattern.

NTSC color-bar generator and a keyed-rainbow generator is in the hues and saturations that are provided. With reference to Fig. 2–20, an NTSC color-bar signal contains the three primary colors and their complementaries, plus a reference white level. These colors are generated at full brightness and saturation. On the other hand, a keyed-rainbow pattern displays an array of mixed colors with variable brightness and saturation. However, this distinction is of minor importance in routine troubleshooting procedures, and the economy of the keyed-rainbow instrument has a dominant appeal in the service field.

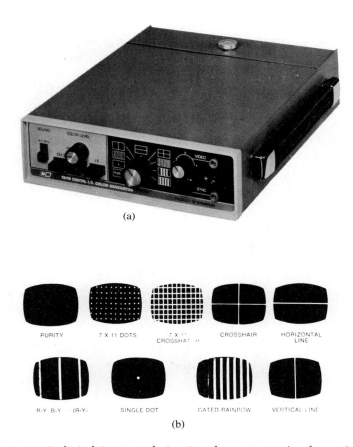

(a)

(b)

Figure 2–25. A digital integrated-circuit color generator/analyzer: **(a)** Appearance of instrument; **(b)** Nine screen patterns available. *(Courtesy of B & K Precision Mfg. Co. Division of Dynascan Corporation)*

When modules are serviced, a bench power supply such as that illustrated in Fig. 2–26 is required. This unit can supply 6 volts at 20 amperes and 12 volts at 10 amperes, and is continuously adjustable over the 0-to-35 volt range. A current demand of 2 amperes is provided on the variable range. The power supply can also be used as a battery charger for 6-volt or 12-volt batteries. Internal protective circuitry is provided, so that the power supply is not damaged if subjected to accidental short circuits. Both voltage and current meters are provided for checking the supply voltage and current demand.

Figure 2–26. A bench power supply. *(Courtesy of Sencore, Inc.)*

With regard to tube-type TV receivers, the tubes are prime suspects in case of receiver malfunction. Accordingly, the first rule is to test or replace the tubes at the outset. Although it is possible for more than one tube to fail completely at the same time, it is not probable. Therefore, when tube trouble occurs, reception can generally be improved by replacement of the "worst" tube. Several weak tubes may also be found, and replacement of these tubes may restore the receiver to normal operation. On the other hand, if receiver operation is still unacceptable, the troubleshooter must proceed to make electrical tests and measurements in the pertinent circuitry. A typical tube tester is illustrated in Fig. 2–27. This instrument provides multiple sockets for rapid plug-in testing of common types of tubes, with a control-set-up section for testing the less common types of tubes. All tubes, except for diodes and rectifiers, are checked for mutual conductance (transconductance). Diodes and rectifiers are emission-tested.

Figure 2–27. Appearance and major features of a typical tube tester. (Courtesy of B & K Precision Mfg. Co. Division of Dynascan Corporation)

REVIEW QUESTIONS

1. How does a color-TV receiver differ from a black-and-white receiver?
2. What is the color-subcarrier frequency?
3. State the highest significant frequency in a horizontal sync pulse (approximately).
4. Name a component defect that could cause misalignment of an RF or IF tuned circuit.
5. Explain the factors in a video amplifier that determine its high-frequency response.

6. Why does a mistimed gating pulse impair color-sync action?
7. How is an oscilloscope applied to analyze color-sync circuit action?
8. Briefly discuss the problem of IF regeneration.
9. What is the basic relation between hue and phase?
10. How are chroma phases usually checked?
11. State a common cause of IF overload.
12. Describe the chief features of a television analyzer.
13. How does an NTSC color-bar generator differ from a keyed-rainbow generator?
14. When trouble symptoms occur in tube-type receivers, what component or device falls under preliminary suspicion?
15. Name the parameters that are customarily checked by a tube tester.

Troubleshooting the Bandpass Amplifier, ACC, and Color-killer Sections

3.1 BANDPASS AMPLIFIER OPERATION

As shown in Fig. 3–1, the bandpass amplifier operates between the black-and-white section and the color detectors. It was noted previously that the chroma section is disabled during reception of a black-and-white signal. On the other hand, the chroma section is enabled when an incoming color signal is sensed. In other words, the bandpass amplifier is automatically switched on or off by the control voltage from the color killer. When the bandpass amplifier is enabled, the automatic chroma control (ACC) section also becomes operative and maintains the chroma-signal output from the bandpass amplifier at a constant amplitude regardless of variations in the amplitude of the incoming signal. The color-killer section is associated with circuitry that responds to the presence or absence of a color burst on the back porch of the horizontal-sync pulse. Unless a color-killer network is provided, color-noise interference, called *confetti,* would be reproduced in the picture during reception of black-and-white programs.

Separation of the chroma signal from the Y signal occurs in the bandpass amplifier, as depicted in Fig. 3–2. Note that the bandpass amplifier has a center frequency of 3.58 MHz, and a passband of ±0.5 MHz. In other words, the bandwidth of the bandpass amplifier is substantially less than the bandwidth of the

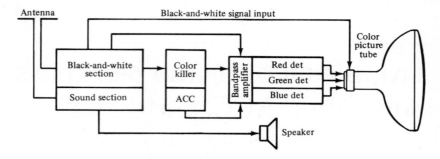

Figure 3–1. Block diagram for the bandpass-amplifier, ACC, and color-killer sections.

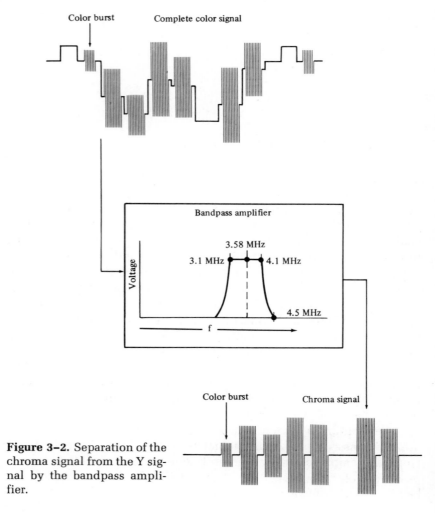

Figure 3–2. Separation of the chroma signal from the Y signal by the bandpass amplifier.

video amplifier. If the bandwidth becomes abnormal because of a component defect, hue distortion occurs, owing to interference by the black-and-white signal. On the other hand, if the bandpass amplifier has subnormal bandwidth owing to a component defect, color reproduction becomes smeared and color disappears from the smaller areas of the picture. Therefore, when this type of color-picture trouble symptom occurs, the technician should check the frequency response of the bandpass amplifier.

Typical picture symptoms resulting from various component defects in the bandpass-amplifier section are weak hue reproduction or no color, distorted hues, poor color "fit," and/or abnormally intense color reproduction. Color "fit" denotes the relative positioning of the black-and-white image and the color image. As an illustration, the color "fit" is not normal if the lipstick is shifted horizontally from an actress' lips, and reproduced a quarter of an inch from its normal position. A block diagram showing the basic interrelations of the bandpass amplifier, color-killer, and automatic chroma control sections is given in Fig. 3–3. The burst amplifier is essentially a narrow-band 3.58-MHz tuned stage that turns the ACC and color-killer sections on or off. Unless a burst signal is present, there is no output from the burst amplifier. In this example, a two-stage bandpass amplifier is utilized; the ACC control voltage is applied to the bandpass-amplifier driver, and

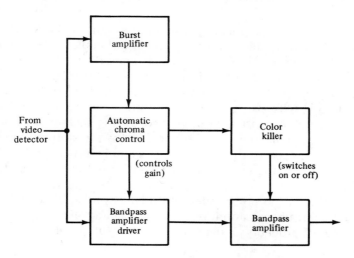

Figure 3–3. Block diagram for the bandpass amplifier, color-killer, and ACC sections.

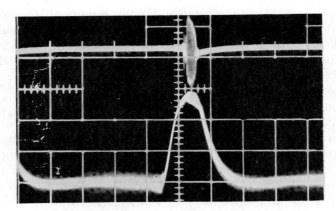

Figure 3–4. Normal waveforms in the burst-amplifier section. *(Courtesy of Sencore, Inc.)*

the color-killer control voltage switches the second stage of the amplifier on or off. Normal waveforms in the burst-amplifier section are shown in Fig. 3–4.

3 . 2   CIRCUIT ACTION IN BANDPASS-AMPLIFIER AND COLOR-KILLER SECTIONS

Troubleshooting in the bandpass-amplifier and color-killer sections requires an understanding of the circuit actions that occur in the associated networks. Figure 3–5 *(see pages 60 and 61)* shows the circuitry for a typical color-killer, bandpass-amplifier, and color-sync arrangement. At this time, we are not concerned with the color-sync network. Observe the 250-k color-killer control at the bottom of the diagram. If confetti is displayed during reception of black-and-white programs, the color-killer control is probably set too low. However, if the color killer is out of range, it is indicated that there is a defective component in the color-killer network. Conversely, if a color image is not displayed when the receiver is tuned to a color-TV signal, the color-killer control has probably been set too high. If a color image cannot be displayed at any setting of the control, there is probably a defective component in the color-killer section.

Although trouble symptoms can be caused by various kinds of component and device defects, leaky or open capacitors are the most common troublemakers. As an illustration, with reference to Fig. 3–5, if the 150-pf capacitor to the base of Q3 is "open," or if the 0.047-µf capacitor to the base of Q4 is "open," the color killer cannot be activated by a color-burst signal. Again, if the 3300-pf capacitor at the collector of Q9 becomes leaky or "shorted," confetti will be displayed on the picture-tube screen regardless of the setting of the color-killer control. A typical oscilloscope display of the color burst and gating pulse was shown in Fig. 3–4, and the NTSC color-bar display is seen in Fig. 3–6. Open capacitors can usually be pinpointed to best advantage by checking the associated waveforms. As an example, if a color burst is observed at the emitter of Q3, but the burst waveform is absent at the base of Q4, it would be concluded that the 0.047-µf capacitor is "open."

Leaky or "shorted" capacitors can usually be pinpointed by DC-voltage measurements. In doubtful situations, resistance measurements with a hi-lo ohmmeter can often provide helpful test data. When color reproduction is obtained on strong incoming signals only, the input waveforms to the chroma sections should be checked. In case an input waveform is weak or absent, the trouble is not in the chroma section, but in a preceding black-and-white section. On the other hand, if the input waveform has normal amplitude and is not distorted, as specified in the receiver service data, the trouble will be found in the chroma section(s). After the general trouble area has been localized, the technician proceeds to

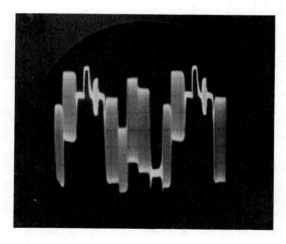

Figure 3–6. An NTSC color-bar waveform, as displayed on an oscilloscope screen.

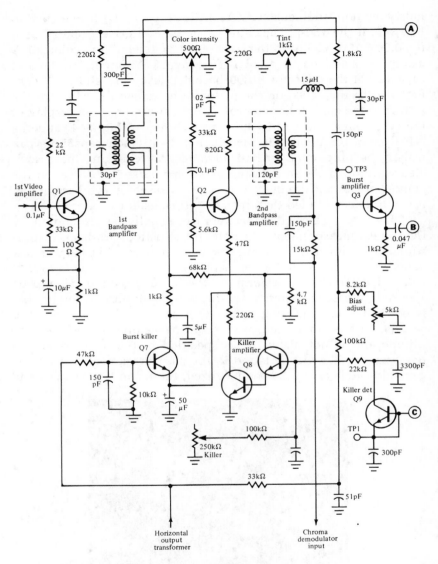

Figure 3–5. Configuration for a typical color-killer, bandpass amplifier, and color-sync arrangement.

check the DC voltages at the transistor terminals. Voltage measurements may also be made at other circuit points.

In Fig. 3–7, two series of DC voltages are specified. Thus, the

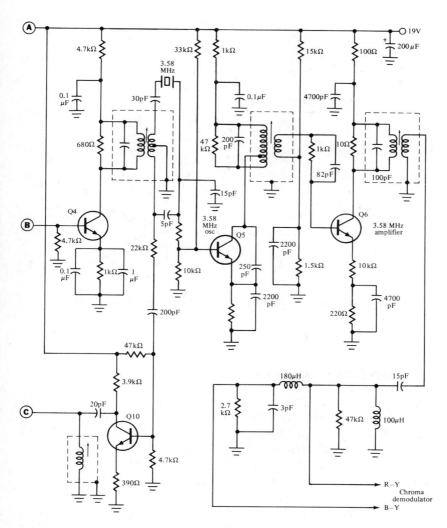

Figure 3–5. Continued

base and emitter voltages for the transistor are specified for color reception and for black-and-white reception. Observe that the transistor is reverse-biased by 1.4 volts during black-and-white reception. On the other hand, it is forward-biased by 0.3 volt during color reception. Similarly, the color-killer source voltage is 70 volts during black-and-white reception, but rises to 225 volts dur-

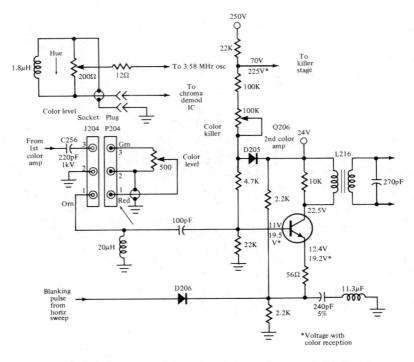

Figure 3–7. Bandpass-amplifier circuitry, with specified DC voltages.

ing color reception. This change in DC-voltage levels is not always noted in receiver service data. However, when voltages are specified both for signal and no-signal conditions, the technician's task becomes easier. In other words, if an expected voltage change does not occur, the additional test data thus provided are often of considerable assistance in pinpointing the defective component or device.

When DC-voltage measurements are inconclusive, supplementary resistance measurements are generally made. Resistance checks are always made in circuit, if possible, to avoid the time and effort required to unsolder and resolder connections. In-circuit resistance measurements require observance of the circuit arrangement, and may also require analysis of the test results. It is instructive to consider the basic factors that are involved. With reference to Fig. 3–8, Q1 will normally be cut off provided that its collector is negative. Therefore, to measure the value of R1 in circuit with a conventional ohmmeter, the test leads must be polar-

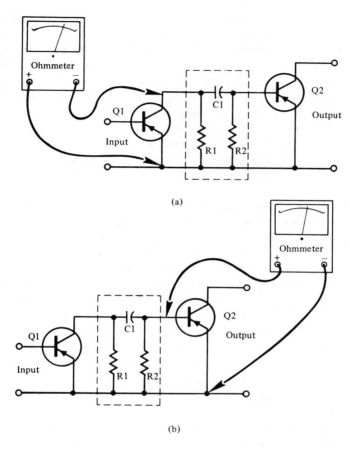

(a)

(b)

Figure 3–8. Ohmmeter test leads must be polarized as shown to avoid "turning on" transistor junctions: **(a)** Measurement of R1 resistance; **(b)** Measurement of R2 resistance.

ized as shown. Again, Q2 will normally be cut off, provided that its base is positive. Therefore, to measure the value of R2 in circuit with a conventional ohmmeter, the test leads must be polarized as shown. Otherwise, the low shunt resistance of a "turned-on" junction would make the resistance reading meaningless.

If a hi-lo ohmmeter is used on its low-power ohms function to make the foregoing resistance measurements, test-lead polarity can be disregarded. In other words, the lo-pwr ohmmeter applies less than 0.1 volt between the circuit points under test. In turn, the junctions of a normal transistor cannot be "turned on" during resistance measurements. There are various circuits that are un-suitable for in-circuit resistance measurements with a conven-tional ohmmeter. As an illustration, consider the measurement of the resistance of R3 in circuit with the configuration of Fig. 3–9. The ohmmeter test-lead polarity must necessarily apply a negative voltage to D2 and a positive voltage to D1, but then the collector of Q1 becomes positive, and the collector junction is "turned on." In this situation, the resistance of R3 must be measured with a hi-lo ohmmeter operated on its lo-pwr ohms function.

Note that the in-circuit resistance measurement of R3 in Fig. 3–9 may be incorrect, even if a hi-lo ohmmeter is used. In other words, D1, D2, or Q1 may have leaky junctions. In such a case, the resistance of R3 would falsely appear to be subnormal. To re-solve the problem, one end of R3 can be disconnected from its cir-cuit, so that an out-of-circuit resistance measurement may be made. It is not absolutely necessary to unsolder the connection of R3 to its circuit board. Some technicians prefer to make a razor cut across the printed-circuit conductor to the resistor. This opens the resistor circuit, and an out-of-circuit resistance measurement can be made. Then, the printed-circuit conductor is repaired by melting a small drop of solder across the razor cut.

There are two basic in-circuit transistor tests that can be made in many configurations, which provide a quick check on

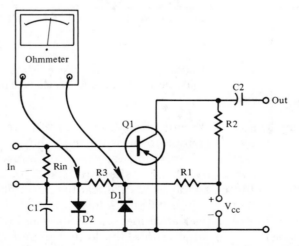

Figure 3–9. The resistance of R3 cannot be measured accurately with a conventional ohmmeter.

transistor workability. These are called the *turn-off test* and the *turn-on test*. A turn-off test is made as depicted in Fig. 3–10. With a DC voltmeter connected between collector and ground, a temporary short circuit is applied between the base and emitter of the transistor. If the transistor has normal control action, its collector voltage jumps up to V_{cc}. This turn-off test is based on the fact that a normal bipolar transistor cuts off when its base and emitter are brought to the same potential. In turn, collector-current flow through R_L is stopped, and the collector voltage rises to the supply-voltage value. If the collector voltage does not increase to V_{cc}, it is indicated that the transistor has a leaky collector junction, and should be replaced.

Next, a turn-on test is made as shown in Fig. 3–11. A DC voltmeter is connected across the emitter resistor, and the forward bias on the transistor is temporarily increased by connecting a 50-k bleeder resistor between the collector and base terminals. A normally operating transistor responds by drawing more emitter current. In turn, the voltage drop across R_E increases when the turn-on test is made. When bias-stabilized transistor configurations are tested, as exemplified in Fig. 3–12, the turn-on and turn-off tests noted above are less useful, because of side effects. In

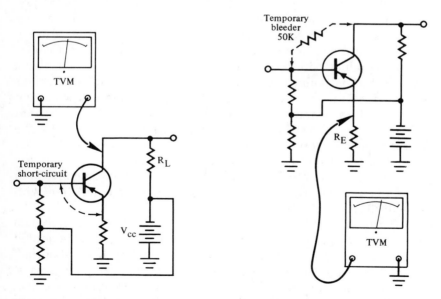

Figure 3–10. Transistor turn-off test. **Figure 3–11.** Transistor turn-on test.

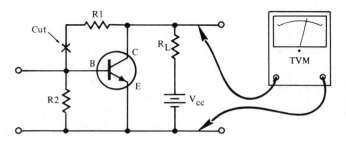

Figure 3–12. Another basic turn-off test.

other words, R1 bleeds collector current in a conventional turn-off test, and a temporary bleeder resistor places an extra current demand on R_L. Thus, both of these conditions tend to mask the test results and make them more or less inconclusive. Therefore, it is helpful in this situation to make a turn-off test by slitting the PC conductor to R1, as indicated in Fig. 3–12. Then, if the collector voltage jumps up to the V_{cc} value, it is indicated that the transistor is responding normally. Finally, a small drop of solder is melted over the razor slit to repair the PC conductor.

Another type of turn-off test must be made in tuned-amplifier stages that have an inductive collector-load configuration, as exemplified in Fig. 3–13. Since the winding resistance of T2 is very low, practically no DC-voltage drop occurs across the primary winding. However, a turn-off test can be made without difficulty. The PC conductor in the primary circuit is slit with a razor blade, and a TVM is connected across the ends of the conductors as shown. In turn, the TVM is operated on a suitably low-current range, and the meter indicates the quiescent collector-current value. A turn-off test is made by temporarily short-circuiting the base and emitter terminals of the transistor, as shown in the diagram. If the collector-current reading drops to practically zero, the transistor is responding normally. Otherwise, it is defective and should be replaced. Finally, the slit PC conductor is repaired with a small drop of solder.

Turn-off tests often provide helpful supplementary data when DC-voltage measurements are inconclusive. In RC-coupled stages, for example, a transistor may be found to have abnormal forward bias and to be drawing too much current. In such a case, DC-voltage measurements do not necessarily pinpoint the defective component. Thus, the coupling capacitor may be leaky, or a

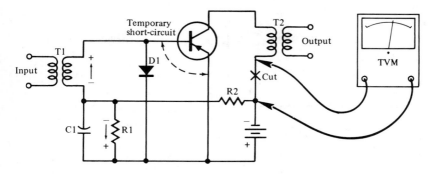

Figure 3–13. Turn-off test for inductive collector-load configuration.

resistor may be off-value, or the transistor may have excessive leakage. In turn, it is advisable to make a turn-off test to determine whether the transistor is defective. Then, if the test shows that the transistor is not at fault, the technician is one step nearer to pinpointing the defective component.

3 . 3 SERVICING TUBE-TYPE CHROMA CIRCUITRY

It is instructive to observe the typical tube-type bandpass-amplifier and color-killer configuration shown in Fig. 3–14. Although the circuit functions are essentially the same as in a corresponding solid-state arrangement, there are various technical differences that are of importance to the troubleshooter. When malfunction occurs, tubes are suspected first. Note that the DC voltages through the tube network are much higher than in a comparable transistor configuration. Both pentode and triode tube types are utilized in the example of Fig. 3–14, whereas triode transistors are used exclusively in its solid-state counterpart. In-circuit resistance measurements can be made with a conventional ohmmeter in the case of tube circuitry, because the tubes are effectively open circuits when no power is applied. In many cases, a resistance chart such as that shown in Fig. 3–15 is included in the receiver service data. Resistance charts facilitate in-circuit resistance measurements, because they eliminate the need for calculation.

Direct-current-voltage values are specified on the circuit diagram in most cases, as seen in Fig. 3–14. These are "bogie" values,

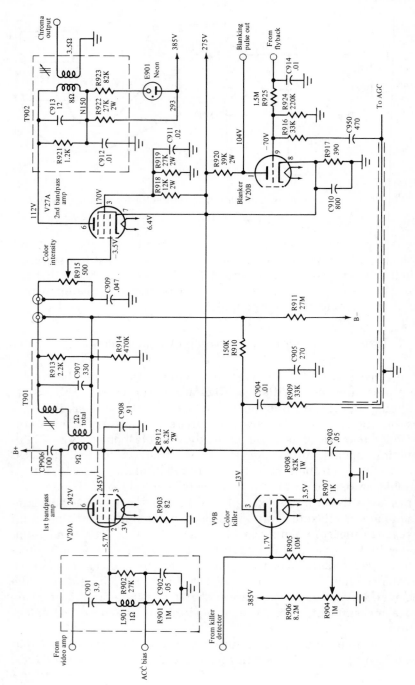

Figure 3–14. A tube-type bandpass-amplifier and color-killer configuration.

RESISTANCE MEASUREMENTS

ITEM	TUBE	PIN 1	PIN 2	PIN 3	PIN 4	PIN 5	PIN 6	PIN 7	PIN 8	PIN 9	PIN 10	PIN 11	PIN 12	TOP CAP
V1	11AF9	4700Ω	330Ω	57Ω ††	55K #†	12.2Ω	11.2Ω	460Ω	1000Ω	12.5K †	4100Ω †			
V2	21LU8	8.2Ω	5.8M ‡	NC	240Ω †	NC	2.4M	2.4M	22K †	1150Ω	100K	150K	11.2Ω	
V3	6LX8/LCF802	10K †	120K	11.5K †	7.5Ω	8.2Ω	30K †	0Ω	1000Ω	1.1M				
V4	40KD6	13Ω	0Ω	13K †	0Ω	500K	NC	NC	NC	NC	0Ω	NC	14Ω	4.1Ω ‡
V5	34CE3	0Ω	NC	NC	24Ω †	NC	NC	500K	NC	NC	24Ω †	NC	7.5Ω	
V6	3BT2				Pins 1 thru 12 have infinite resistance									550Ω †
V7	8AC10	12.2Ω	39K †	190Ω	190Ω	39K †	190Ω	10K	0Ω	10K	39K †	10K	13Ω	
V8	15XP22	FIL	4200Ω †	350K †	1.5M †	1.7M †	5600Ω †	350K †	NC	0Ω	NC	5400Ω † PIN 13 1.7M †	350K † PIN 14 FIL	
ITEM	TUBE	PIN 1	PIN 2	PIN 3	PIN 4	PIN 5	PIN 6	PIN 7	PIN 8	PIN 9	PIN 10	PIN 11	PIN 12	TOP CAP

This reading will vary depending upon the condition of the electrolytic in the circuit.
‡ Measured from pin 7 of V5.
† Measured from cathode of X2.
†† Measured from cathodes of X4 and X5.
NC No connection

Figure 3–15. A typical resistance chart. (Redrawn by permission of Howard W. Sams & Company, Inc.)

which are subject to a permissible tolerance of ±20 percent in most cases. As an illustration, the plate voltage for V20A is specified at 242 volts. In practice, the technician would regard a reading from 194 to 290 volts as acceptable. On the other hand, a reading outside of this range would throw suspicion upon the plate circuit of the tube; it could also throw suspicion upon the grid circuit or upon the screen circuit, in case the grid (or cathode) voltage or the screen voltage were incorrect. Note that the plate circuit for V27A includes a neon bulb; in case the plate voltage is incorrect, this neon bulb should be checked. A neon bulb, like an electron tube, has a comparatively limited life expectancy, although tubes are more likely to fail than are neon bulbs.

If all tubes are found to be in good condition (or defective tubes have been replaced), capacitors are the next most likely suspects. Paper capacitors become defective more often than mica or ceramic capacitors. Leakage is the most common fault, although a defective capacitor may also be found to be open-circuited. Leaky or short-circuited capacitors are often pinpointed by means of DC-voltage measurements. For example, if C91 in Fig. 3–14 were leaky, the plate and screen voltages on V20A would be subnormal. On the other hand, if C909 were quite leaky, this fault would not affect the grid voltage on V27A noticeably. If the capacitor were practically short-circuited, the technician could detect this defect by means of a coil-resistance measurement (the specified resistance of the coil is 2 ohms). Note that substantial leakage in C907 will have the effect of reducing the signal output from T901, and will make the tuned-circuit response excessively broad. This latter malfunction can be determined by means of a frequency-response check.

A basic rule to keep in mind is that malfunction is never assumed to be the result of misalignment, unless it is known that the set owner has tampered with the alignment adjustments. When substantial misalignment occurs in the usual course of operation, it is the result of a component failure. As an illustration, if C913 in Fig. 3–14 happens to become "open," T902 will be thrown out of alignment. Open-circuited capacitors can often be quickly checked by a "bridging" test. For example, if the color-intensity control does not change the color intensity and the color reproduction has abnormal intensity, it would be suspected that C909 may be "open." In turn, if a known good capacitor of the same value were temporarily shunted across the terminals of C909, and R915 resumed normal action, it would be evident that C909 had become "open" and must be replaced.

In practice, it is sometimes difficult to correlate the chassis components with the schematic symbols. For example, the technician may not be able definitely to identify the bandpass-amplifier tube on the chassis, particularly if the marking on the tube is partly or wholly obliterated. Again, the tubes may have been removed from their sockets, and the technician may be in doubt concerning their correct locations. In these and various other situations, a chassis-layout diagram such as that shown in Fig. 3–16 is most helpful. It reveals at a glance where each major component is located with respect to others. Chassis-layout diagrams are provided in all receiver service data. Note in passing that a seemingly incorrect type of tube will be found in a socket occasionally. However, reference to a tube-substitution manual may show that the particular type of tube is a satisfactory substitute for the type specified in the chassis-layout diagram.

Some tube-type receivers employ a series heater network, as exemplified in Fig. 3–17. Accordingly, an open heater in any tube will cause the receiver to be inoperative. Locating the tube with the open heater involves systematic replacement of the tubes in the "dead" string until the heaters start to glow. If the heater voltage is subnormal, suspicion falls upon the dropping resistor(s), such as R1. In other words, a resistance measurement should be made to determine whether the resistor may have increased in value. RF chokes such as L1, L2, and L3 can also cause disturbance of heater voltages, although this fault is less common than resistor defects. Other causes of heater-circuit trouble are poor socket contacts and cold-soldered joints. Note that if a desoldering tool is used to disconnect a defective component, the used solder should be discarded. When solder is reused, it is likely to make a defective bond owing to oxidation.

3 . 4 ACC CIRCUIT TROUBLESHOOTING

Next, observe the ACC configuration associated with Q1 in Fig. 3–18. Color-picture trouble symptoms that can result from malfunction in the ACC section include abnormally high color in-

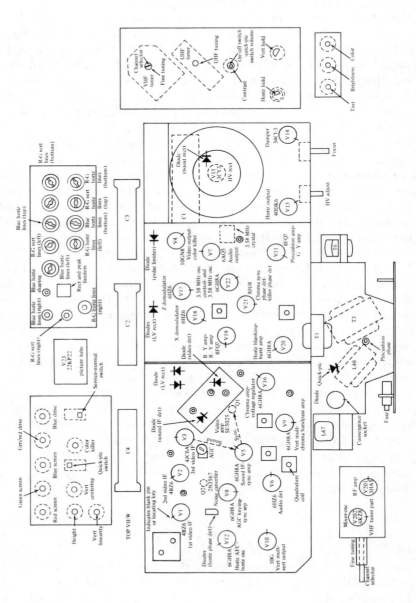

Figure 3–16. A typical tube-type chassis layout. (Redrawn by permission of Howard W. Sams & Company, Inc.)

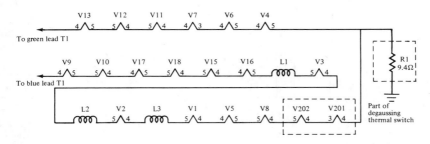

Figure 3–17. Example of a series heater network. *(Redrawn by permission of Howard W. Sams & Company, Inc.)*

tensity, subnormal color intensity, and drifting or intermittent color reproduction. DC-voltage measurements, supplemented by resistance measurements with a hi-lo ohmmeter, plus turn-off or turn-on tests, are employed to close in on defective components or devices. Note that an oscilloscope cannot provide test data in this section, because the ACC network is a DC system. Typical causes of abnormally high color intensity include leaky capacitors, such as the 100-pf coupling capacitor in the D1 ACC diode circuit, an "open" ACC delay diode, a transistor with excessive leakage current, or an off-value resistor. If an ACC transistor is replaced, it is necessary to use a replacement that has closely similar operating characteristics. Otherwise, the resulting ACC action is likely to be unsatisfactory.

To check the operating stability of the ACC section, the technician clamps the ACC output voltage with batteries, or with a bias box. Then, if color reproduction is stabilized, he concludes that the trouble will probably be found in the ACC section. Component defects that cause drifting or intermittent color reproduction are basically marginal faults in the same components that cause abnormal or subnormal color intensity. As an illustration, if a capacitor is leaky, and its leakage resistance tends to drift, color reproduction can be similarly affected. Possible causes of subnormal color intensity include a poor front-to-back ratio in the ACC rectifier D1, leaky capacitors in the base circuit of Q1, a poor front-to-back ratio in the ACC delay diode D2, or a marginally defective ACC-amplifier transistor Q1.

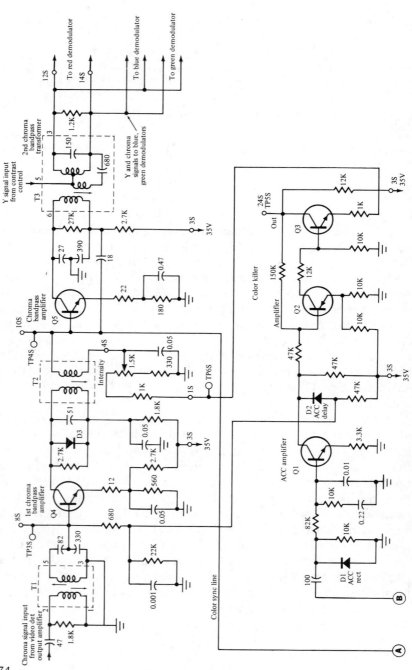

Figure 3–18. Configuration for a typical ACC, bandpass-amplifier, and color-killer arrangement.

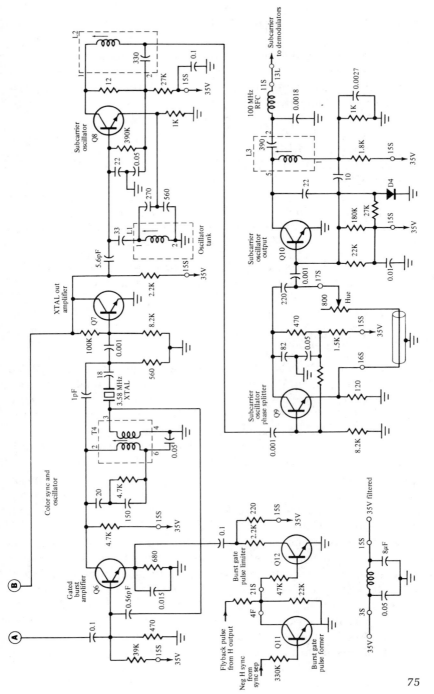

Figure 3-18. Continued

3 . 5   TUBE-TYPE ACC CIRCUITRY

Troubleshooting tube-type ACC circuitry involves most of the basic principles employed in solid-state circuitry. However, there are important differences in detail. A typical tube-type ACC, bandpass-amplifier, burst-amplifier, and color-killer configuration is shown in Fig. 3–19. In case of malfunction, tubes are tested or replaced at the outset. Note that the ACC and color-killer sections utilize the same detector tube. The chroma signal is coupled from the plate of VI via C419 to the grid of V2. In turn, the output from V2 flows through the primary of T402, which is tuned to 3.58 MHz. Accordingly, the gated-out burst signal is coupled from the secondary of T402 via C448 and C449 to the plate and cathode of V3. Simultaneously, the burst signal is mixed with the 3.58-MHz CW output from the sub-carrier oscillator. ACC action takes place as follows:

1. When there is no incoming color burst, the DC voltage at point P in Figure 3–19 is zero. Similarly, there is zero DC voltage at point J.
2. Next, when an incoming color burst arrives, the phase relations in V3 are such that a negative DC voltage appears at points P and J.

Note that this negative bias voltage applied to the grid of V5 changes the bias on V4, thereby turning the bandpass amplifier "on." Also, the negative bias applied to V1 reduces the stage gain, but does not cut V1 off. This bias voltage is proportional to the color-burst amplitude, and thereby maintains the gain of V1 essentially constant under conditions of varying chroma-signal amplitude. In case of malfunction, preliminary trouble localization can generally be made on the basis of oscilloscope waveform checks. In other words, the burst signal can be traced through the network up to the ACC detector. After the trouble area has been localized, DC-voltage measurements often serve to pinpoint the defective component. As explained previously for solid-state ACC tests, it is often helpful to clamp the ACC line. As an illustration, negative DC voltages from adjustable bias boxes may be applied at points P and J to establish desired operating levels while waveforms are being traced and the DC-voltage distribution is being checked.

Service data for chroma circuitry sometimes specify both active and inactive DC voltages. In other words, DC voltages are specified for conditions of chroma signal present, and of chroma signal absent.

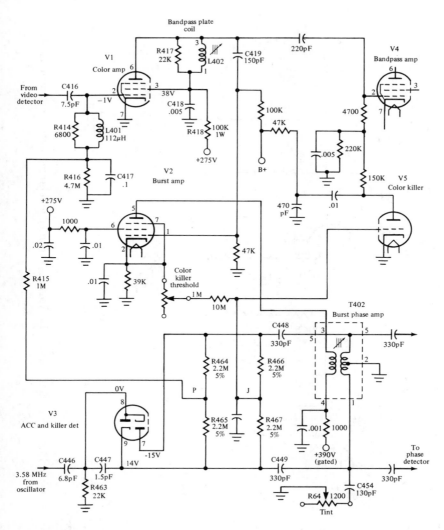

Figure 3–19. A typical tube-type ACC, bandpass-amplifier, burst-amplifier, and color-killer configuration.

As an illustration, Fig. 3–20 shows a schematic diagram for a tube-type bandpass amplifier with DC-voltage values for color signal present, and for color signal absent. Note that the grid of the first bandpass-amplifier tube is biased at −0.6 volt with color signal absent, but is biased at −7 volts with color signal present. Also, the plate and

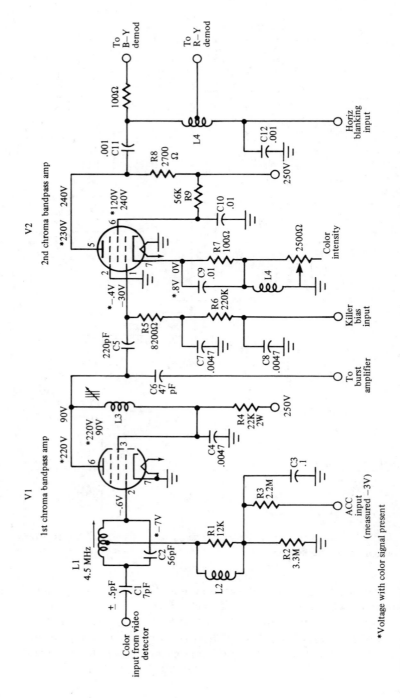

Figure 3–20. Tube-type bandpass-amplifier configuration, with DC voltages specified for chroma signal present and absent.

screen voltages on V1 change from 90 volts to 220 volts when a color signal arrives. The second bandpass-amplifier tube operates in the opposite way; in other words, the grid voltage on V2 is −30 volts with color signal absent, but is only −0.4 volt with color signal present. Also, the plate voltage on V2 changes from 240 volts to 230 volts when a color signal arrives. Similarly, the screen voltage on V2 changes from 240 volts to 120 volts when a color signal arrives. Observe also that the cathode voltage on V2 changes from 0 volt to 0.8 volt when a color signal arrives.

Specification of DC-voltage values under both signal and no-signal conditions can be very helpful in practical troubleshooting procedures. Preliminary evaluation is made by checking the DC voltages at the various tube electrodes while an incoming color signal is switched off and on. If the specified voltage change does not take place at a particular terminal, the technician knows that a circuit fault will be found in that general area. Dual DC-voltage specifications are also helpful in sectionalizing a chroma system in case the trouble symptoms do not point to particular stage malfunction. Thus, if the technician is in doubt whether the malfunction is in the ACC, the color-killer, or the bandpass-amplifier section, dual DC-voltage specification provides him with a more comprehensive error pattern from which to reason the logical cause for a given trouble symptom.

3 . 6   CHROMA WAVEFORM ANALYSIS

Troubleshooting in the bandpass amplifier and following sections often involves chroma waveform analysis. In turn, it is instructive to consider the basic principles that are involved. Waveform analysis is concerned with amplitude, waveshape (distortion), frequency, and phase. As an illustration, the burst keying pulse has a specified peak-to-peak voltage, a certain width (duration), a typical shape, a frequency of 15,750 Hz, and a phase that is normally coincident with the color-burst interval. A specified peak-to-peak voltage value is subject to a reasonable tolerance, just as DC-voltage values have permissible tolerances. As a general rule, the peak-to-peak voltage specified for a waveform will have a permissible tolerance of ±20 percent. For example, the three burst keying-pulse displays depicted in Fig. 3–21 are regarded as falling within the normal range. On the other hand, if a pulse were 30

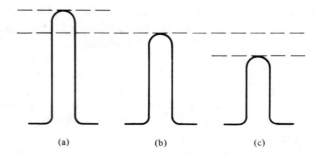

Figure 3–21. Example of normal waveform amplitude tolerance: **(a)** Upper limit; **(b)** Specified value; **(c)** Lower limit.

percent greater or 30 percent less than the specified amplitude, it would be regarded with suspicion, and the technician would investigate the cause of the excessive departure from normal amplitude.

Tolerances on waveshapes often require experienced judgment on the part of the technician. This judgment is based chiefly upon a good knowledge of the circuit action that is involved. For example, consider the three pulse widths depicted in Fig. 3–22. The abnormal pulse width shown at (a) is undesirable in a burst-keying pulse, because it permits some of the chroma signal following the burst to enter the burst amplifier. In turn, the burst signal becomes more or less contaminated with chroma information, with the result that the hues tend to drift back and forth with the changing chroma signal. By way of comparison, the pulse width shown at (b) is the specified width; it permits all of the color burst to pass into the burst amplifier, but prevents any preceding or following signal information from entering. Next, the subnormal pulse width exemplified at (c) is very objectionable, because it permits only a fraction of the burst to pass. In turn, color sync becomes unstable, particularly on weak-signal reception.

When the color burst is displayed by a triggered-sweep scope, it has the typical appearance shown in Fig. 3–23. The important considerations are that the burst waveform have the peak-to-peak voltage specified at the particular test point, and that the burst be complete. In other words, the burst waveform should have eight or nine cycles, although the cycles at the beginning and at the end of the burst will have more or less reduced amplitude.

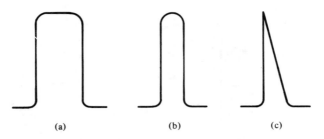

(a)　　　　　(b)　　　　　(c)

Figure 3–22. Pulse width variations: **(a)** Abnormal; **(b)** Specified; **(c)** Subnormal.

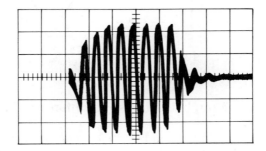

Figure 3–23. Appearance of the color burst, displayed on the screen of a wide-band triggered-sweep scope.

If, for example, only two or three cycles were displayed in the burst waveform, the burst gate would fall under suspicion. A check of the gating waveform might show, for example, that the gating pulse has a subnormal width, such as depicted in Fig. 3–22(c).

When one is measuring waveform amplitudes, it is important to recognize the relations among rms, peak, and peak-to-peak voltage values. With reference to Fig. 3–24, a sine wave has an rms voltage that is 0.707 of its peak voltage. In turn, the peak-to-peak voltage of a sine wave is 1.414 times its rms voltage. Thus, a sine wave that has an rms voltage of 1 volt has a peak voltage of 1.414 volts, and has a peak-to-peak voltage of 2.83 volts. In other words, the peak-to-peak voltage of a sine wave is equal to 2.83 times its rms voltage. Note that conventional AC voltmeters indicate the rms voltage of sine waves. Most service-type meters have both rms and peak-to-peak scales; the peak-to-peak indication will be cor-

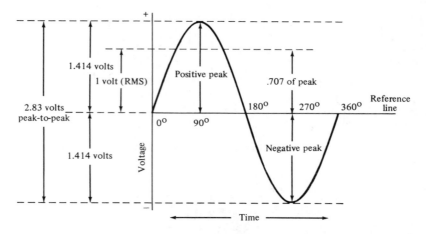

Figure 3–24. RMS, peak, and peak-to-peak voltages in a sine waveform.

rect for any waveform, but the rms indication will be correct only for sine waves.

Chroma waveforms often have two or more components. As an illustration, the keyed-rainbow waveform shown in Fig. 3–25 has a sync-pulse component and a chroma-bar component. The peak-to-peak voltage of the sync pulse is normally equal to the peak-to-peak voltage of the chroma component. The complete color signal shown in Fig. 3–2 has four components; these are the horizontal sync pulse,

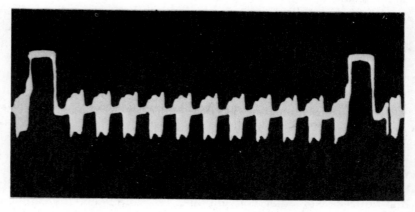

Figure 3–25. Keyed-rainbow waveform has a sync-pulse component and a chroma component.

the color burst, the chroma signal, and the Y signal. Each component has a specified amplitude; thus, the sync pulse normally has 25 percent of the amplitude of the Y signal, the color burst normally has the same amplitude as the sync pulse, and the chroma signal normally has the relative excursions shown in the diagram. Each of the chroma signals has a certain phase that can be checked from related waveforms in the chroma section of the receiver. This topic is detailed subsequently.

Waveforms may be either positive-going or negative-going, as exemplified in Fig. 3–26. If a waveform is passed through a common-emitter stage, its polarity becomes inverted. On the other hand, a waveform passed through a common-collector stage or through a common-base stage has an output polarity that is the same as its input polarity. Most oscilloscopes are designed so that a positive-going excursion extends upward on the screen, whereas a negative-going excursion extends downward on the screen. However, this is not an invariable rule. Therefore, when the technician is using an unfamiliar scope, he should check its display polarity, to determine whether a positive-going sync pulse, for example, will be displayed as in Fig. 3–26(a), or as in (b). This is sometimes an essential consideration, as when one is checking circuit operation after diodes have been replaced. If a diode is accidentally connected in reverse polarity, the associated output waveform will

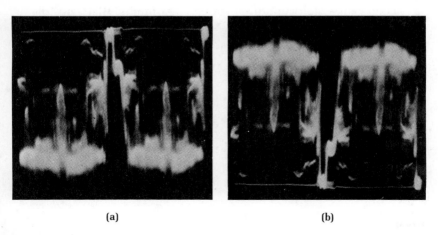

(a)   (b)

Figure 3–26. Composite video signal, displayed at line-rate deflection: **(a)** Positive-going signal; **(b)** Negative-going signal.

also have reverse polarity, and serious malfunction is usually the result.

REVIEW QUESTIONS

1. How is the bandpass amplifier enabled and disabled?
2. Explain the chief function of the bandpass amplifier.
3. Define color "fit".
4. What is the correct setting for the color-killer control?
5. Discuss the localization of leaky capacitors.
6. Why is it helpful to specify DC-voltage values both for color reception and for black-and-white reception?
7. Briefly describe in-circuit resistance measurement procedures.
8. What is a transistor *turn-off* test? A *turn-on* test?
9. When may turn-off or turn-on tests be particularly informative?
10. How does tube-type chroma circuitry differ from solid-state chroma circuitry?
11. Is chroma-circuitry malfunction generally assumed to result from misalignment?
12. Name some color-picture trouble symptoms that can result from malfunction in the ACC section.
13. Why is an oscilloscope often helpful in ACC troubleshooting procedures?
14. With what factors is chroma waveform analysis concerned?
15. Briefly discuss tolerance considerations on waveshapes.

Color-sync Troubleshooting Procedures

4.1 GENERAL CONSIDERATIONS

Trouble analysis in the color-sync system requires an understanding of the circuit actions in this network. There are two basic types of subcarrier-oscillator configurations. The simplest arrangement consists of a 3.58-MHz quartz crystal that is shock-excited into oscillation by the color burst, as depicted in Fig. 4–1. In other words, when a color burst is applied to the crystal, it is shock-excited into a transient oscillation. The amplitude of the output waveform decays, but another transient waveform starts as the next color burst is applied to the crystal. The output waveform in Fig. 4–1 indicates normal operation. Oscillator malfunction is indicated by rapid waveform decay. Rapid decay is caused by abnormal loading in the oscillator circuit, such as by a leaky capacitor.

A more elaborate subcarrier-oscillator arrangement employs a crystal-controlled free-running 3.58-MHz oscillator, which is locked to the frequency and phase of the color burst by an automatic phase control (APC) subsection. Figure 4–2 depicts the plan of this method for subcarrier regeneration. Note that a shock-excited oscillator does not require APC circuitry, because the ringing crystal automatically follows the phase of the burst signal. Both of the foregoing subcarrier-oscillator arrangements have their advan-

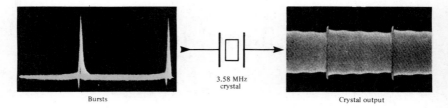

Bursts   Crystal output

Figure 4–1. Principle of a ringing-crystal subcarrier oscillator.

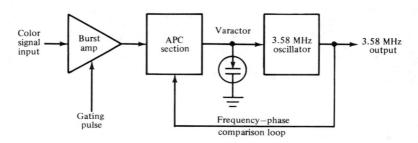

Figure 4–2. Plan of the APC type of frequency-phase controlled free-running subcarrier oscillator.

tages and disadvantages. A shock-excited oscillator has fewer components to become defective, but it does not maintain as tight color synchronization when the incoming signal is comparatively weak. Therefore, the free-running oscillator with APC is employed in the majority of color receivers. Table 4–1 provides a summary of this color-sync system.

A block diagram for a ringing-crystal subcarrier-oscillator section is shown in Fig. 4–3. From the gated burst amplifier, the color-burst voltage is applied to a 3.58-MHz quartz crystal. A limiting amplifier is provided to clip the peaks of the transient waveform, and thereby to produce an output that has a uniform amplitude. The output from the limiter is called the *regenerated color subcarrier*. It is fed to a phase-shift network that develops a two-phase output, as depicted in Fig. 4–3(b). These outputs have the R-Y and B-Y phases, in this example. Note that the limiter output actuates the color killer, which in turn enables or disables the bandpass amplifier (not shown in Fig. 4–3). In other words, unless there is an incoming burst signal, there is no output from the limiter and no control-voltage output from the color-killer to enable the bandpass amplifier.

TABLE 4–1

Summary of Subcarrier-Oscillator/APC Color-Sync System

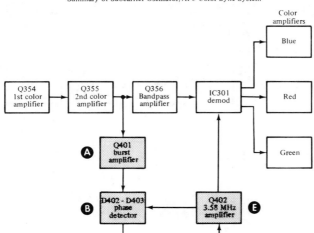

A. Amplifies the 3.58 MHz "burst" signal that is transmitted by the TV station. This signal is used as a reference for the demodulators to determine the intensity of each beam. The beam intensity, in turn, determines the proper amount of each color.
B. Compares the output frequency of the 3.58 MHz reference signal oscillator with the burst frequency, and generates a correction voltage based on the comparison.
C. Changes capacitance as the correction voltage applied to it changes. This corrects the 3.58 MHz oscillator and makes it the same frequency as the burst signal.
D. Creates a continuous 3.58 MHz reference voltage; provides the reference to the demodulators to obtain correct color signals.
E. Amplifies the 3.58 MHz reference signal.

Next, consider the operation of an automatic frequency-phase controlled (AFPC) free-running subcarrier oscillator, as depicted in Fig. 4–2. This arrangement is also called an APC configuration. Note that the output from the burst amplifier is fed to the APC section, which is essentially a discriminator, as exemplified in Fig. 4–4. In other words, the APC circuit compares the burst frequency and phase with that of the 3.58-MHz oscillator output. If the oscillator output tends to lead or lag the burst phase, the discriminator responds by developing a positive or negative DC control voltage. This control voltage is applied to a varactor diode, which responds with a change in effective capacitance. This varactor diode is connected in parallel with the 3.58-MHz quartz crystal, as seen in Fig. 4–5. Since the resonant frequency of a quartz crystal can be changed over a limited range by shunt-capacitance variation, the end result is to pull the crystal oscillator back in phase with the color burst.

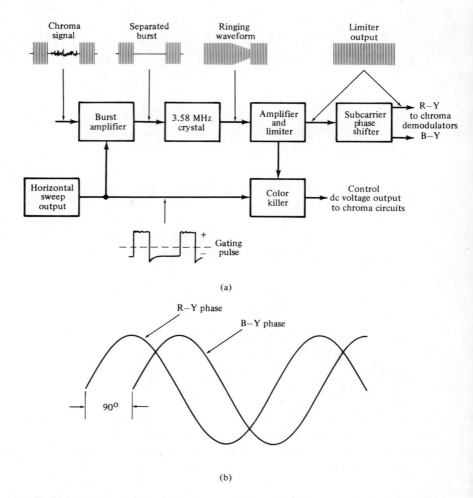

Figure 4–3. Block diagram for a ringing-crystal color-subcarrier oscillator section: **(a)** Functional subsections; **(b)** Two-phase output to chroma demodulators.

4 . 2   BURST, APC, AND SUBCARRIER-OSCILLATOR
        CIRCUIT BOARD

Next, it is instructive to observe the configuration for a burst, APC, and subcarrier-oscillator circuit board, shown in Fig. 4–6. The 3.58-MHz subcarrier oscillator, IC401, operates in a Colpitts

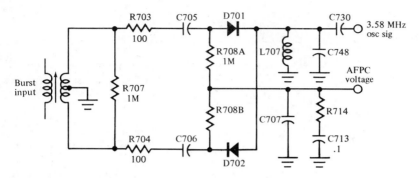

Figure 4–4. The APC (AFPC) section is essentially a discriminator.

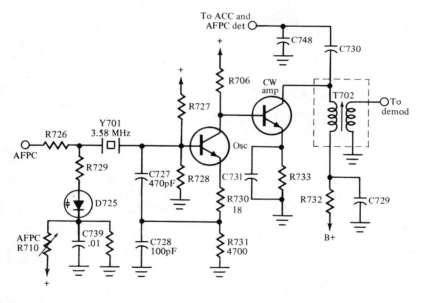

Figure 4–5. Skeleton circuit diagram for the APC and subcarrier oscillator.

circuit. This integrated circuit functions as an amplifier and has the internal circuitry depicted in Fig. 4–7. The feedback energy required to sustain oscillation of the quartz crystal is coupled from output terminal 7 of IC1 to input terminal 3 by the capacitance voltage divider C15 and C13. Note that the frequency of oscillation is determined by the resonant frequency of the quartz crystal CR1 as modified by the capacitance of varactor diode D6. In turn, the

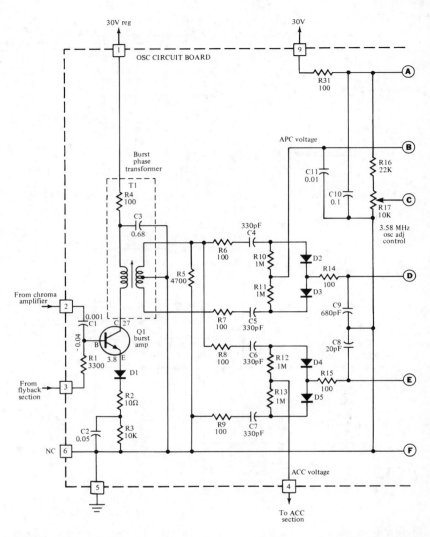

Figure 4–6. Configuration for a burst-amplifier, APC, and subcarrier-oscillator circuit board.

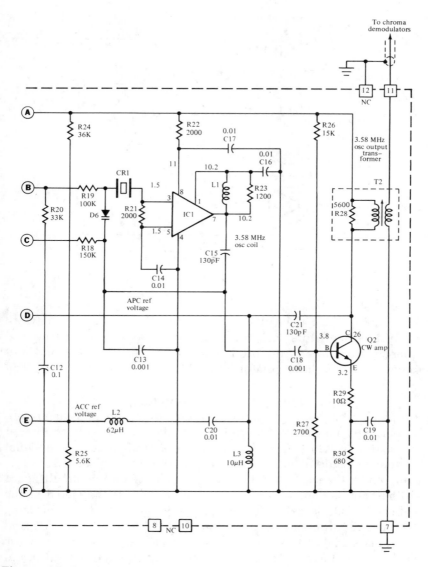

Figure 4–6. Continued

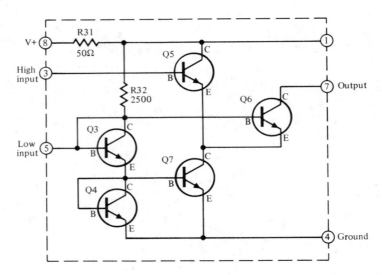

Figure 4–7. Internal circuitry of the IC shown in Fig. 4–6.

APC voltage and the voltage established by the divider network comprising resistors R16 and R17 serve to determine the frequency and phase of the oscillator output. Resistors R18 and R19 serve to isolate the oscillator-input AC signal from the DC control voltages.

The output from the oscillator circuit in Fig. 4–6 is then coupled through C18 to the base of Q2, which operates as an amplifier. Base bias for this stage is established by the voltage divider consisting of R26 and R27. In turn, the amplified reference signal at the collector of Q2 flows through the primary of T2 and is coupled by the secondary through terminal 11 on the circuit board to the chroma-demodulator section. A portion of the signal at the collector of Q2 is fed back through the voltage divider comprising C21 and C9 through R14 to the junction of the APC diodes D2 and D3. This signal at the junction of the capacitance-divider network is shifted 90 degrees in phase by the network comprising C9 and C20 with coils L2 and L3. This signal then feeds through R15 to the junction of the ACC diodes D4 and D5.

Note that the chroma signal from the bandpass (chroma) amplifier is coupled via C1 to the base of Q1 in Fig. 4–6. Transistor Q1 operates as a gated burst amplifier with a positive-going gating pulse applied through R1. In turn, the gated-out burst voltage drops across the primary of T1. During conduction, R3 sets the correct operating point for the emitter of Q1, and C2 provides by-pass action. Resistor R2 introduces sufficient emitter degeneration for adequate operating stability and also establishes the stage gain. Cutoff bias for Q1 between successive bursts is supplied by the discharging of C2 through R3. In other words, the transistor is reverse-biased by signal-developed base voltage. Diode D1 is in-cluded to prevent this reverse bias from exceeding the reverse emitter-breakdown voltage of the transistor. Correct bandwidth for T1 is obtained by connection of the 4700-ohm resistor R5 across its secondary terminals. Normal bandwidth for the burst amplifier is depicted in Fig. 4–8. Note that the burst-amplifier bandwidth is substantially less than the bandpass-amplifier bandwidth.

A center tap is provided for the secondary of T1 in Fig. 4–6 so that a two-phase output is provided with equal 3.58-MHz volt-ages that are 180 degrees out of phase with each other. One phase is fed through R6 and C4 to the anode of D2; the other phase is coupled through R7 and C5 to the cathode of D3. In turn, when the burst signal and the reference signal from the subcarrier os-cillator are compared in the diodes, an output voltage will be pro-duced if there is any difference in frequency or phase between the two signals. Any resulting DC output voltage will appear at the junction of R10 and R11. The polarity of this control voltage will be positive or negative, depending upon the sign of the phase or

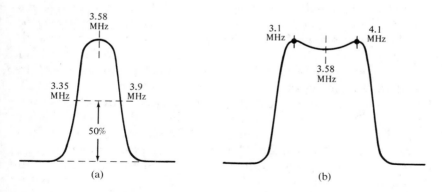

Figure 4–8. Normal bandwidth for the burst-amplifier stage: **(a)** Burst-amplifier frequency response curve; **(b)** Bandpass-amplifier response curve (for comparison).

frequency error, while its amplitude will depend upon the magnitude of the error. This control voltage is in turn applied to the anode of varactor diode D6, which produces a corresponding frequency or phase correction in the subcarrier output voltage. Note that the APC control voltage is filtered into pure DC by C11 and R20 with C12.

Observe that the burst signal in Fig. 4–6 is also supplied through R8 and R9 with C6 and C7 to the ACC diodes D4 and D5. The ACC detector operates in the same manner as that of the APC detector, except that the 3.58-MHz reference input is shifted 90 degrees in phase. This results in a negative DC output voltage at the junction of R12 and R13, with an amplitude that is proportional to the amplitude of the burst signal. Also, a fixed DC bias voltage, which is provided via R24 and R25, is applied through D5 to this junction. This fixed DC voltage and the prevailing ACC voltage are then fed from terminal 4 on the circuit board to the chroma circuit board.

4.3   TROUBLESHOOTING PROCEDURES

Analysis of color-sync trouble symptoms usually starts with measurement of DC voltages. For example, the voltages at the transistor and IC terminals would be compared with specified values, as exemplified in Fig. 4–6. It is good procedure also to check the supply voltage to the oscillator circuit board. If this voltage is substantially subnormal, color sync will be lost, and the DC-voltage distribution of the entire network will be affected. A fault may or may not become obvious after the DC voltages have been checked. For example, a diode with a poor front-to-back ratio can result in unstable color-sync action, without causing a serious upset in specified DC voltage values. Again, an open capacitor such as C11 in Fig. 4–6 produces confusing changes in DC circuit voltages.

It is often desired to make a test that will indicate whether color-sync trouble is located in the APC section or in the subcarrier-oscillator section. This test is made as shown in Fig. 4–9. The control voltage to the varactor is clamped by the output from a variable bias box. As the clamp voltage is slowly varied, it may be observed that the color image can be free-wheeled into sync. At this point, it is known that the oscillator can operate on frequency. However, the technician still does not know whether

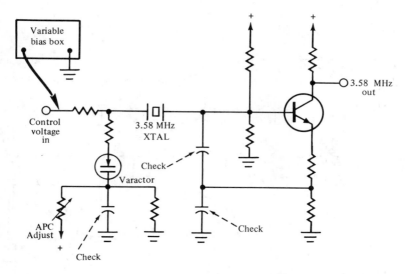

Figure 4–9. APC-oscillator response may be analyzed with a variable bias source.

the trouble is in the oscillator section or the APC section. There-fore, the DC voltages are now measured in the oscillator section. If these voltages are normal, it is concluded that the fault is in the APC section. On the other hand, if the oscillator-circuit voltages are incorrect, it is indicated that there is a fault in the oscillator section. The fixed capacitors in the configuration are prime sus-pects; they should be systematically checked for leakage and "opens."

Color-sync malfunction is sometimes caused by a defective burst transformer. However, since the associated components such as capacitors and diodes are more likely to become defective, these should be checked out first. Then, if it appears that the burst transformer may be defective, a frequency-response check should be made, as shown in Fig. 4–10. A signal generator is coupled to the primary circuit by means of a gimmick; this consists of a turn or two of insulated wire around the primary lead to the trans-former. A transistor voltmeter (TVM) is connected at the APC out-put, to indicate the voltage developed by the APC rectifier diodes. The subcarrier oscillator is disabled by any convenient means, such as unplugging the 3.58-MHz crystal. Then the signal-generator frequency is varied through the range from 3 to 4 MHz.

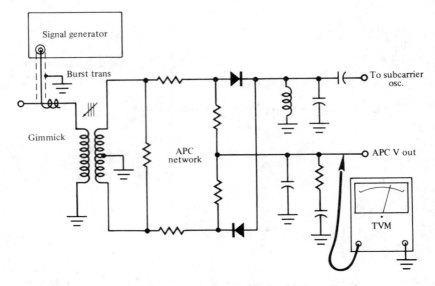

Figure 4–10. Frequency-response check of burst transformer.

Peak response should be indicated on the TVM at 3.58 MHz. If necessary, the slug adjustment should be touched up. Then, the response is checked at 3.35 MHz and 3.9 MHz (Fig. 4–8), to determine whether the output is 6 dB (50 percent) down, as specified. Excessive bandwidth and/or inability to peak the transformer at 3.58 MHz indicates that there is an internal defect such as shorted turns. In such a case, the transformer should be replaced.

In color-program reception, the color-burst frequency is 3.579545 MHz, which is commonly rounded off to 3.58 MHz. On the other hand, when the color-sync system is operated from a keyed-rainbow generator, the color-burst frequency is 3.563795 MHz, commonly rounded off to 3.56 MHz. Although there is a 15,750-Hz difference between these two burst frequencies, the color-sync section normally has adequate pull-in range to lock on a keyed-rainbow signal. In case the subcarrier oscillator locks in on a color-program signal, but breaks sync lock on a keyed-rainbow signal, it is indicated that synchronizing action is too critical, and that there is a component defect in the APC network or in the subcarrier-oscillator circuit. However, before measuring DC voltages or making sectional checks, it is advisable in this situation to try varying the APC adjust control (Fig. 4–9).

Obsolete tube-equipped color-oscillator circuits with a reactance stage controlled by a diode-type phase detector were usually optimized for 3.58-MHz operation by grounding the output of the phase detector (APC detector), thereby simulating a normal zero control voltage, and then adjusting the reactance coil until an upright color image floated (freewheeled) horizontally on the screen. This procedure was called "zero-beating the oscillator." However, the solid-state equivalent of this arrangement, exemplified in Fig. 4–11, has an important difference in that forward bias is employed. Thus, pin 2 of IC92 has approximately +1.25 volts present, from inside the IC. Therefore, the output of the phase detector should not be grounded during adjustment of the reactance coil. Instead, the base of the burst amplifier M108 should be grounded. Then the slug in L100 may be precisely adjusted for zero beat. Figure 4–12 shows the appearance of a typical chroma board.

The foregoing procedure is also useful to determine whether loss of color sync is caused by a weak burst signal, or by off-frequency operation of the subcarrier oscillator. Thus, when the base of M108 is grounded, the number of color stripes or "rainbows" in the out-of-sync image provides a useful clue. If there are less than two or three color stripes, poor color-sync action is probably being caused by an attenuated burst signal. On the other hand, if there are many color stripes in the image, it is indicated that the subcarrier oscillator is operating considerably off frequency. In turn, the technician looks for a defective component or device in the oscillator section or in the APC section. In case of doubt, as when three color stripes are displayed in the image, it is advisable to check the burst waveform with a scope, and compare its peak-to-peak voltage with the value specified in the receiver service data.

It is instructive to note that the sequence of hues in an out-of-sync color image shows whether the subcarrier oscillator is running too fast or too slow. Thus, with reference to Fig. 4–13, as we look down any one column, the sequence of colors is green-blue-red-green-blue-red. This sequence indicates that the subcarrier oscillator is operating at a frequency above 3.58 MHz. On the other hand, if the sequence of colors as we look down a column were green-red-blue-green-red-blue, the subcarrier oscillator would be operating at a frequency below 3.58 MHz. Also, the number of rainbows displayed from the top to the bottom of the screen indicates the value of the frequency error. In other words, if only

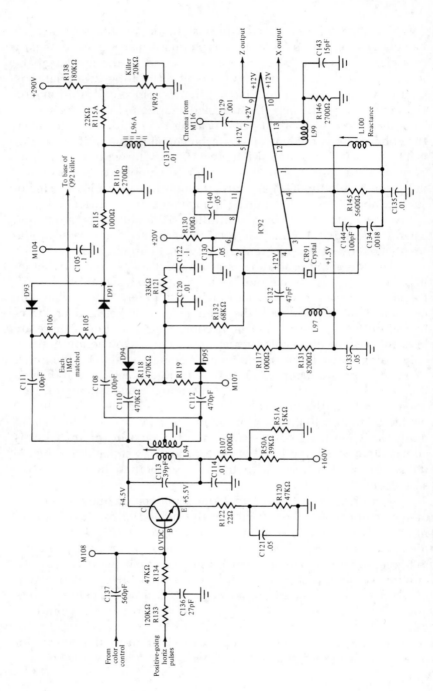

Figure 4–11. Reactance coil is adjusted with M108 test point grounded.

Figure 4–12. Appearance of a typical chroma circuit board. *(Courtesy of Heath Company)*

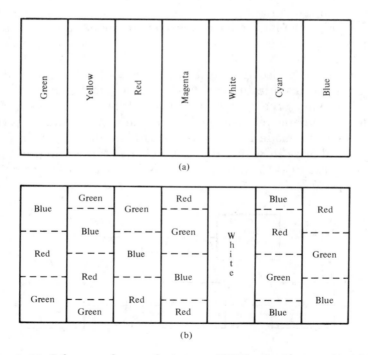

(a)

(b)

Figure 4–13. Color sync-loss analysis in an NTSC color-bar pattern: **(a)** Pattern in color sync; **(b)** Pattern with loss of color sync.

one rainbow is displayed, the frequency error is 60 Hz. If two rainbows are displayed, the frequency error is 120 Hz, and so on.

When there is no color reproduction, and it is suspected that the subcarrier oscillator is "dead," a scope is generally used to check for a 3.58-MHz output waveform from the oscillator. Sometimes a scope is not available, and it is desired to make another type of test. A useful signal-injection test is depicted in Fig. 4–14. This test utilizes the 3.58-MHz output from a signal generator to substitute for the quartz-crystal source. In other words, the generator output cable is connected across the terminals of the crystal socket. If the no-color symptom results from an inoperative oscillator, an out-of-sync color image is likely to be displayed on the screen when the receiver is tuned to a color signal. Of course, this test may be inconclusive under some circumstances. As an illustration, if the oscillator transistor happens to be short-circuited, the generator signal cannot pass into the chroma-demodulator section, and no color display will be obtained. If a color display appears, however, it is confirmed that the oscillator is "dead."

Troubleshooting is sometimes hampered by lack of service data for a particular model of receiver. In such situations, it may be possible to make comparative DC-voltage measurements and waveform checks in a similar receiver that is in good operating condition. Note also that different models of receivers produced by the same manufacturer generally have some sections that are the same, and other sections that are different. In this circumstance, comparative DC-voltage measurements and waveform checks can be made in sections that are similar. Inevitably, situations will arise in which the technician is thrown entirely upon his own resources. In turn, he must fall back upon his knowledge

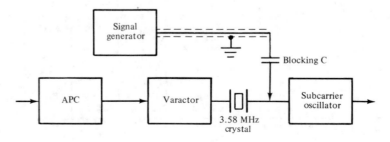

Figure 4–14. Signal-injection test in the subcarrier-oscillator section.

of circuit action and his experience with component characteristics.

Many modern color receivers have modular construction. Although this feature does not make troubleshooting easier, it greatly reduces the down time and permits practically all repairs to be completed in the home. A suspected module is quickly checked by merely plugging in a replacement module. Modules with faults can be tested and repaired subsequently in the shop at any convenient time. Circuit boards have the same basic advantage, although a suspected circuit board cannot be replaced as quickly as a module. In other words, various connecting leads must be unsoldered and resoldered when a circuit board is replaced. For this reason, the technician usually prefers to take a receiver to the shop if a circuit board needs to be replaced.

4 . 4   TUBE-TYPE COLOR-SYNC SYSTEM

Troubleshooting a tube-type color-sync system such as that depicted in Fig. 4–15 starts with tube replacement. If the trouble symptom(s) must be investigated further, DC-voltage measurements are generally made. It is also informative to check the color-burst signal with an oscilloscope at the input of the chroma phase detector. For example, the burst signal may not be arriving at the phase detector. Further localization can be made by applying a variable DC control voltage to the chroma-reference oscillator-control tube, at point F. Note that the normal grid-bias voltage for V2 is 0.2 volt in this example. In other words, when the grid of the tube is clamped by the output from a bias box, and the bias voltage is slowly varied through 0.2 volt, the color picture will normally "freewheel" into sync. If color sync cannot be temporarily obtained by clamping, it is logical to conclude that the trouble will be found in the V2 or V3 circuitry.

Most circuit malfunctions are the result of leaky, short-circuited, or open-circuited capacitors. Electrolytic capacitors fail oftener than paper capacitors, and paper capacitors fail more often than mica or ceramic capacitors. A leaky capacitor can often be pinpointed by means of DC-voltage tests, although this is not an invariable rule. Leaky capacitors may show up on in-circuit resistance measurements, if the leakage is substantial. Open capacitors can seldom be pinpointed by DC-voltage measurements, except in

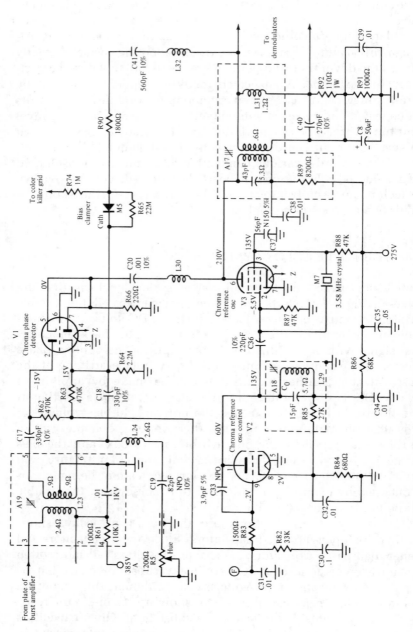

Figure 4-15. Typical tube-type color-sync system.

the case of a coupling capacitor in a stage that operates with signal-developed bias. When a capacitor is suspected of being open-circuited, the technician usually makes a "bridging" test by temporarily shunting the suspect with a known good capacitor, to observe whether normal operation is resumed.

An oscilloscope check can be made to determine whether there is 3.58-MHz output from the chroma reference oscillator, in case of doubt. If the oscillator is "dead," C_o in Fig. 4–15 is a ready suspect. In other words, if C_o is leaky or short-circuited, the circuit will not oscillate. Sometimes the quartz crystal becomes defective; if this trouble is suspected, a substitution test should be made. Note that the grid bias of —5.5 volts on V2 is a signal-developed bias. Thus, this DC voltage is a good indicator of oscillation; when the oscillator is "dead," the grid bias is practically zero. Note that if C36 becomes leaky, the oscillator will not function, and a small positive bias will be measured on the grid of V2. In the event that a defective capacitor is not the cause of color-sync trouble symptoms, semiconductor diodes and resistors are checked next. For example, M5 may be found to have a poor front-to-back ratio. Variable resistors such as R5 are more likely to deteriorate and cause circuit malfunction than are fixed resistors. A worn and "noisy" potentiometer generally causes trouble by drifting in value.

4 . 5   SIGNAL GENERATOR CHARACTERISTICS

Various types of signal generators are utilized in color-TV troubleshooting. A typical service-type generator is illustrated in Fig. 4–16. This instrument has a frequency range from 100 kHz to 30 MHz, with a rated accuracy on frequency calibration of ±3 percent. RF output voltage is adjustable from 5 to 100,000 μV, or 0.1 volt. Amplitude-modulated output is available with variable percentage up to 50 percent. Technicians often prefer the higher accuracy and maximum output provided by laboratory-type test oscillators, such as that shown in Fig. 4–17. This instrument has a frequency range from 10 Hz to 10 MHz, and is provided with an output meter to indicate how many microvolts or volts are being applied to the point of test. A maximum output of 3.16 volts can be supplied, and its minimum output is practically zero. Be-

Figure 4–16. A general-purpose service-type signal generator. (*Courtesy of EICO*)

Figure 4–17. A laboratory-type test oscillator (*Courtesy of Heath Company*)

cause this type of generator has a comparatively high output level, signal-injection tests in chroma circuitry are considerably facilitated.

4 . 6   AUXILIARY TEST-EQUIPMENT UNITS

Auxiliary test-equipment units provide substantial convenience in practical color-TV servicing procedures. For example, the variable bias-voltage supply shown in Fig. 4–18 is a valuable time-saver in bench work. It is applied in AGC clamping, ACC clamping, alignment procedures, and other test work where an adjustable low-impedance DC source is needed. This type of bias supply provides either positive or negative outputs from three separate 25-volt sources. Also, one of the sources has an additional switch position that provides up to 75 volts for tests on chroma amplifiers that operate at 67.5 volts.

An RC substitution box, such as that illustrated in Fig. 4–19, is another useful incidental unit. It contains 24 values of resistors from 10 ohms to 5.6 megohms, 9 values of capacitors from 100 pF to 0.5 μF at 600 volts, electrolytic capacitors with values of 10 μF and 100 μF at 450 volts, and 1000 μF at 75 volts, with a built-in surge-protection circuit to prevent excessive transient voltages in

Figure 4–18. A variable bias-voltage supply (*Courtesy of Sencore, Inc.*)

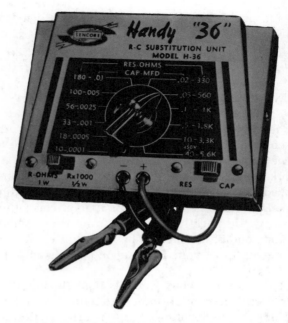

Figure 4–19. An RC substitution box. *(Courtesy of Sencore, Inc.)*

solid-state circuits. It is provided with test leads for connection
into the circuit under test. The chief advantage of an RC substitution box is in the time that is saved in selecting a particular
value of resistance or capacitance for trial in a circuit.

Another auxiliary unit of test equipment used in most service
shops is a capacitor checker, such as that shown in Fig. 4–20. This
instrument measures capacitance values from 10 pF to 1000 μF,
resistance values from 5 ohms to 50 megohms, and tests for leakage at 16 different DC voltages. Terminals are provided for use of
an external standard in combination with the internal Wheatstone
bridge for making supplementary inductance, capacitance, or resistance measurements. It may also be employed in combination
with an external audio oscillator for checking components at frequencies up to 10 kHz. The internal test frequency is 60 Hz. A
comparison-bridge facility is provided which quickly checks turns
ratios on power transformers or other transformers that have ap-
preciable inductance in their windings.

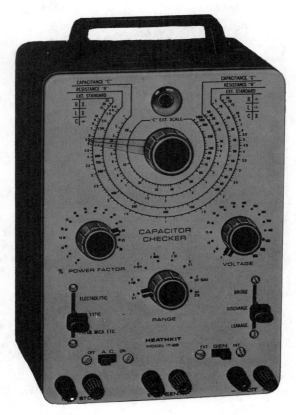

Figure 4–20. A service-type capacitor checker. *(Courtesy of Heath Company)*

Technicians often prefer a specialized vectorscope-chroma generator instrument, as illustrated in Fig. 4–21, instead of a separate generator and scope with vectorscope input facilities. This instrument is comparatively compact, and employs a 3-inch CRT. It contains a standard 10 chroma-bar generator with supplementary white-dot, crosshatch, and shading-bar outputs. Vectorgram patterns are utilized to indicate missing or weak colors, adjustment of the burst-phase transformer, subcarrier oscillator, reactance coil, bandpass transformer, and chroma-demodulation phases. These topics are explained in greater detail subsequently.

Figure 4–21. A combination vectorscope-chroma generator. *(Courtesy of Heath Company)*

REVIEW QUESTIONS

1. What are the two basic types of subcarrier-oscillator configurations?

2. Define a *regenerated color subcarrier.*

3. How does a varactor diode change the frequency of the subcarrier oscillator?

4. Does a burst amplifier have greater or lesser bandwidth than a bandpass amplifier?

5. Explain how color-sync trouble symptoms are analyzed at the outset.

6. State the peak frequency of the burst-amplifier response curve.

7. Describe the difference in subcarrier frequencies employed in a color-TV transmission and in a keyed-rainbow signal.

8. As a color-subcarrier oscillator drifts farther off-frequency, how does the image on the color picture-tube screen change?

9. Can the output from a signal generator be used to make a substitution test in the subcarrier-oscillator circuit?

10. Briefly discuss how a technician may proceed to analyze receiver malfunctions if receiver service data are unavailable.

11. Define a module.

12. Are variable resistors more likely to deteriorate than fixed resistors?

13. How does a lab-type signal generator differ from a service-type generator?

14. Why is an RC substitution box occasionally useful in troubleshooting procedures?

15. Briefly describe the information provided by a vectorgram.

Chroma-demodulator Troubleshooting

5.1 GENERAL CONSIDERATIONS

Trouble analysis in the chroma-demodulator section depends upon an understanding of the circuit actions that are involved. The fundamental color-signal decoding process is shown in Fig. 5-1. In this example, a red bar signal is being decoded. Three chroma demodulators are employed, with the R-Y signal fed to the red grid, the B-Y signal fed to the blue grid, and the G-Y signal fed to the green grid of the color picture tube. At the same time, the Y signal is applied to all three cathodes in the color picture tube. These signals are square-wave voltage levels at this point. Observe that the R-Y and Y signals add to produce full output from the red gun; the B-Y and Y signals cancel to give zero output from the blue gun; the G-Y and Y signals cancel to give zero output from the green gun. In turn, a red hue is displayed on the screen of the color picture tube.

It is instructive to observe how the R-Y and B-Y signals are decoded from the complete color signal. With reference to Fig. 5-2, the output from the bandpass amplifier is applied to both the R-Y and the B-Y demodulators. Also, outputs from the subcarrier oscillator are fed to both demodulators. One of these 3.58-MHz voltages has the R-Y phase, and the other has the B-Y phase.

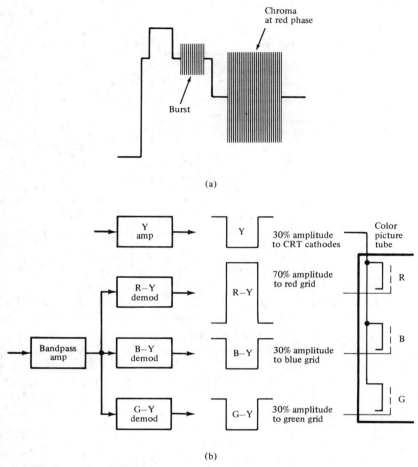

(a)

(b)

1. Y voltage adds to R−Y voltage.
2. Y voltage subtracts from B−Y voltage.
3. Y voltage subtracts from G−Y voltage.

Figure 5-1. Fundamental color-signal decoding process: **(a)** Encoded red color-bar signal; **(b)** Basic decoding action.

Demodulation occurs on the peaks of the injected subcarrier voltage. Thus, the demodulators operate as samplers and develop R-Y and B-Y outputs as shown in Fig. 5-3. In other words, each demodulator normally conducts only for a short interval during the peak of the cycle. Note that as the R-Y signal is going through its peak, the B-Y signal is going through zero.

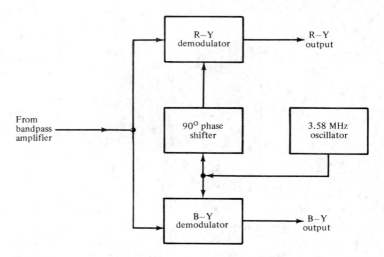

Figure 5–2. Basic block diagram of an R-Y/B-Y demodulator section.

Malfunction of chroma demodulation occurs in case of a component defect that causes the injected 3.58-MHz voltages to have a phase angle other than 90 degrees. Demodulation phase angles can be checked in more than one way. Many technicians prefer a vectorscope check. A basic vectorscope test setup and a typical vectorgram are shown in Fig. 5–4. The vectorscope input terminals are connected at the outputs of the R-Y and B-Y demodulators, and the receiver is connected to a keyed-rainbow generator. In turn, the vertical- and horizontal-gain controls of the vectorscope are adjusted to display the vectorgram pattern with suitable vertical and horizontal proportions. It is helpful to note some basic characteristics of the patterns.

1. If the R-Y demodulator is "dead," only a horizontal line will be displayed on the CRT screen.
2. If the B-Y demodulator is "dead," only a vertical line is displayed on the screen.
3. If color sync is lost, a circular or elliptical pattern will be displayed, as shown in Fig. 5–5.
4. If the injected 3.58-MHz voltages have a 90-degree phase angle, the outline of the vectorgram will be circular.
5. If the injected 3.58-MHz voltages have a phase angle other than 90 degrees, the outline of the vectorgram will be elliptical.

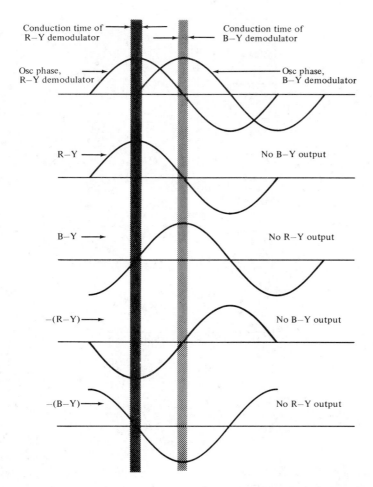

Figure 5–3. Decoding of R-Y and B-Y signals by the chroma demodulators.

It will be observed that the vectorgram will turn clockwise or counterclockwise on the screen as the hue control is adjusted in the receiver. Normally, the vectorgram will be positioned on the screen as seen in Fig. 5–4(b) when the hue control is set to mid-range. Inability to position the vectorgram display properly at any setting of the hue control indicates that there is a component defect in this section. There are also various "fine points" to be observed in vectorgram analysis, as will be explained in following topics.

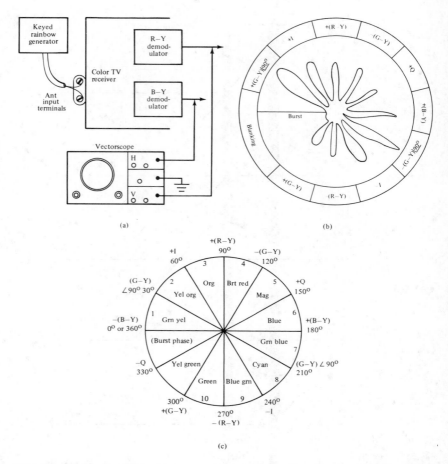

Figure 5–4. Basic vectorscope test: **(a)** Equipment connections; **(b)** A typical vectorgram; **(c)** Phase angles and corresponding colors.

5 . 2   TROUBLESHOOTING CHROMA-DEMODULATOR CIRCUITRY

Demodulation angles are indicated by the vectorgram outline, as depicted in Fig. 5–4. In order clearly to display a vectorgram outline, it is good practice temporarily to disable the color sync section. This can be easily done by short-circuiting the APC output terminal to ground. Then the vectorgram will rotate rapidly on the

Figure 5–5. Vectorscope pattern resulting from loss of color sync.

CRT screen, providing a display such as that shown in Fig. 5–5. By comparing the ellipse thus obtained with the Lissajous figures pictured in Fig. 5–6, the demodulation phase angle can be estimated with sufficient accuracy for practical purposes. Thus, an R-Y/B-Y demodulator system will display a true circle in normal operation (see Fig. 5–7). As will be explained in a following section, other demodulation systems will be encountered in practice.

Incorrect demodulation phase angles are caused by defects in the phase-splitting network between the subcarrier oscillator and the chroma demodulators. Capacitor faults are the most probable, although resistors occasionally become off-value, and inductors may sometimes cause trouble. In the example of Fig. 5–8, the 3-pF and 15-pF capacitors would be checked first for possible leakage or "opens." If the capacitors are satisfactory, the resistors would be checked next. These resistive values are fairly critical for accurate phase shifting action. Finally, if the resistor values are correct, it is indicated that an inductor is faulty. Most service shops are not equipped to measure inductance values. Therefore, the suspected inductors are replaced at the outset.

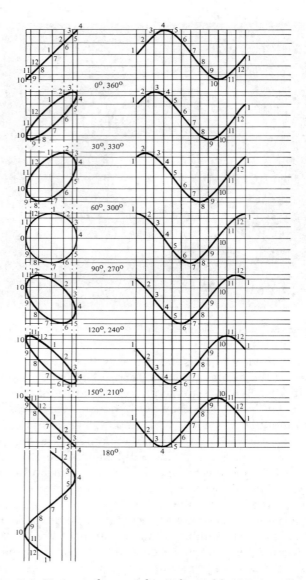

Figure 5–6. Various phase angles indicated by Lissjous figures.

A word of caution is in order concerning ohmmeter checks of inductors. A resistance measurement will indicate whether the coil has continuity, but it cannot indicate the inductance value.

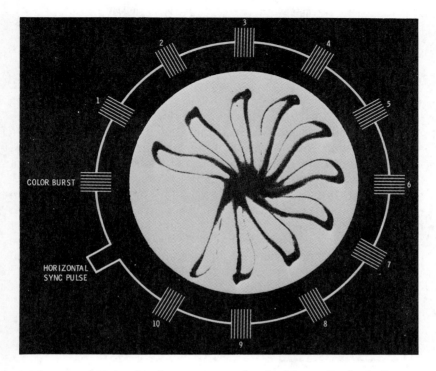

Figure 5–7. Example of a vectorgram that is almost circular in form.

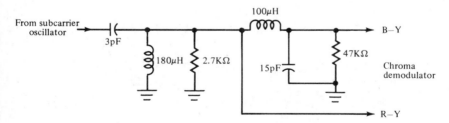

Figure 5–8. Phase-splitting network for chroma demodulators.

Receiver service data may specify the normal resistance of an inductor. Again, this resistance value cannot be employed to select a suitable replacement inductor. In other words, the resistance of a coil depends upon the size of wire with which it is wound. On the other hand, the inductance of a coil depends on the number of turns, the diameter, and the ratio of diameter to length—its inductance is practically independent of the wire size, or winding resistance. Therefore, an exact replacement is required when a phase-shifting network needs a new inductor.

A basic chroma-demodulator configuration is shown in Fig. 5–9. It operates as both a phase detector and an amplitude detector. There is no output from the demodulator unless there is an incoming chroma signal. Again, there is no output from the demodulator if the 3.58-MHz oscillator signal stops, even though an incoming chroma signal is present. The amplitude of the demodulator output depends upon the amplitude of the chroma signal. However, the polarity of the demodulator output depends upon the phase of the chroma signal. There are three points in a demodulator circuit where waveform checks should be made with a scope. These are at the chroma-signal input point, at the subcarrier-injection point, and at the demodulator output point. In case one or more of the waveforms happens to be abnormal, the demodulator diodes should be checked first. An

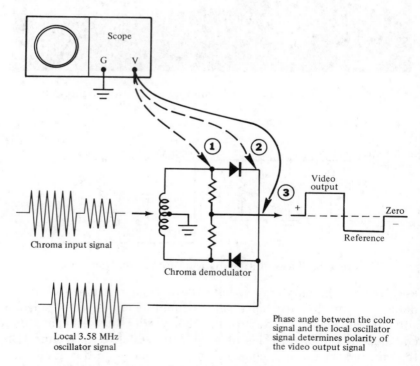

Figure 5–9. Basic chroma-demodulator configuration.

"open," "shorted," or "leaky" (poor front-to-back ratio) diode will result in demodulator malfunction and corresponding waveform distortion.

When a chroma demodulator is driven by a keyed-rainbow signal, the demodulator output waveforms appear as illustrated in Fig. 5–10. Note carefully in this example that the R-Y and B-Y waveforms are acceptable. On the other hand, the G-Y waveform indicates a malfunction in the G-Y channel, owing to the excessive curvature of the base line in the pattern. If this trouble symptom is encountered, the technician should check the bandwidth of the G-Y tuned circuitry. For example, the solid curve in Fig. 5–11 shows an acceptable frequency response for the G-Y demodulator channel. On the other hand, the dotted curve shows a response curve with subnormal bandwidth; this response causes both low- and high-frequency attenuation of the keyed-rainbow signal. High-frequency attenuation produces excessive base-line curvature, as in the G-Y waveform of Fig. 5–10. Low-frequency attenuation pro-

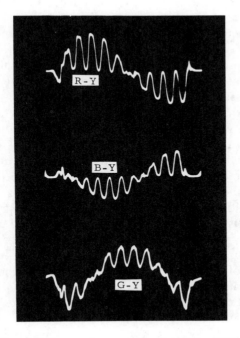

Figure 5–10. R-Y and B-Y waveforms are acceptable; G-Y waveform indicates a circuit malfunction.

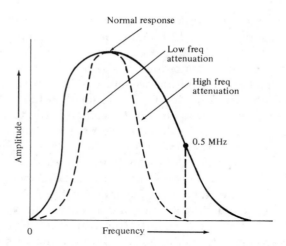

Figure 5–11. Normal and abnormal chroma-demodulator frequency-response curves.

duces the opposite form of base-line curvature, as discussed later in this chapter.

If the demodulation phases are correct, the output waveforms from the R-Y and B-Y demodulators will have the crossovers shown in Fig. 5–12 when the receiver is energized by a keyed-rainbow chroma-bar generator. In other words, the R-Y waveform normally crosses the zero axis on the sixth pulse, and the B-Y waveform normally crosses the zero axis on the third and ninth pulses. These waveforms change in response to adjustment of the hue control. Consider a situation in which the B-Y waveform does not null on its third and ninth pulses when the hue control is adjusted to make the R-Y waveform null on its sixth pulse. This is a symptom of incorrect demodulation phase angles. Conversely, if the R-Y waveform does not null on its sixth pulse when the hue control is adjusted to make the B-Y waveform null on its third and ninth pulses, the technician concludes that the demodulation phase angles are incorrect.

A word of caution is in order when one is checking chroma-demodulator output waveforms in receivers that are provided with automatic tint control (ATC). When the ATC control is partially or fully turned on, the first four pulses in the demodulator output waveforms will be pulled toward, or will be completely merged with, the second pulse. If a vectorgram is displayed with the ATC control turned on, the patterns will be distorted as shown in Fig. 5–13. Although this is the practical way to check ATC action, it

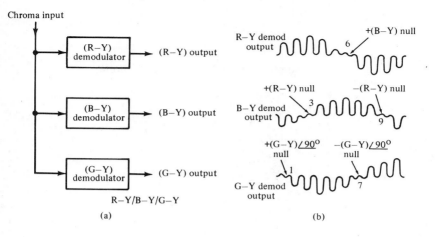

(a)

(b)

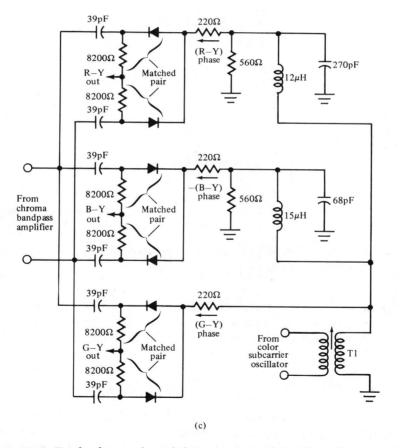

(c)

Figure 5–12. Triple chroma-demodulator arrangment: **(a)** Block diagram of system; **(b)** Normal crossovers of the output waveforms; **(c)** Typical configuration. *(Redrawn by permission of RCA)*

121

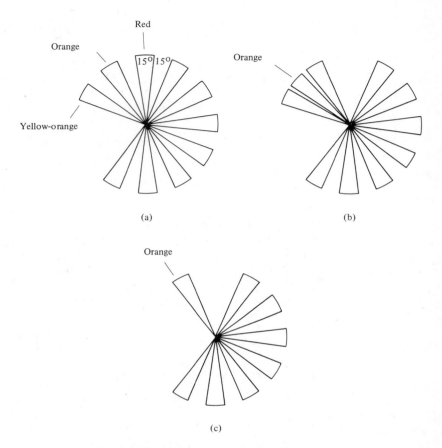

Figure 5–13. Idealized vectorgrams showing ATC action: **(a)** ATC switch off; **(b)** Partial ATC; **(c)** Full ATC.

changes the normal chroma-demodulation angles. Therefore, the ATC control must be turned off when waveforms or vectorgrams are being inspected for chroma demodulation errors.

5 . 3 XZ CHROMA-DEMODULATION SYSTEM

Another common type of chroma-demodulation system is called the XZ system. It employs a demodulation phase-angle difference of 120 degrees, instead of a 90-degree difference, as in the R-Y/B-Y arrangement. This use of 120-degree demodulation axes in the XZ system results in a certain amount of hue distortion. However, the distortion involves a trade-off in that the green-yellow hues are reproduced with an orange cast. In turn, the adjustment of the hue control becomes less critical for reproduction of acceptable flesh tones. Thus, the XZ demodulation system provides some of the advantages of the ATC action depicted in Fig. 5–13. On the other hand, this ATC action is limited to the vicinity of the orange hues, whereas the XZ demodulation system distorts color reproduction in the vicinity of the cyan hues also.

Figure 5–14 shows the plan of the XZ demodulation system. As seen in (a), the chroma-demodulator phases fall 15 degrees beyond the R-Y and B-Y axes. The demodulators are followed by amplifiers and by a G-Y matrix. This matrix is discussed in the next chapter. In Fig. 5–15, the X and Z demodulators are transistors that perform the same basic function of the diodes depicted in Fig. 5–9. In case of malfunction in the demodulator section, some hues may be missing from the color image, whereas other hues are reproduced normally. It is likely that one of the demodulator stages is weak or "dead" in this situation. As an illustration, if the X demodulator becomes inoperative, a keyed-rainbow pattern will be displayed on the picture-tube screen with weak or absent red hues. On the other hand, if the Z demodulator becomes inoperative, a keyed-rainbow pattern will be displayed with weak or absent blue hues.

Scope waveform checks are advisable when demodulator trouble is suspected. Normal keyed-rainbow output waveforms from the X and Z demodulators are shown in Fig. 5–15(b). If one of the waveforms is distorted or absent, the capacitors in the associated circuit should be checked first. For example, C2 and C3 should be checked for "shorts," "opens," or leakage. If the capacitors are not defective, the transistors may be found defective. Observe that turn-off tests can be made in the Q1 and Q2 circuits. In other words, the collector voltage is measured while the base and emitter terminals of the transistor are temporarily short-circuited. If the meter reading jumps up to the supply voltage (+20 volts in this example), it is indicated that the transistor is workable. Note that off-value resistors can also cause trouble symptoms in a chroma-demodulator circuit. Thus, if R6 increases substantially in value, the X and Z waveforms will have incorrect crossovers. Observe also that in case C1 becomes "shorted," there will be no output from Q2.

Vectorgram checks are made in XZ demodulator systems in

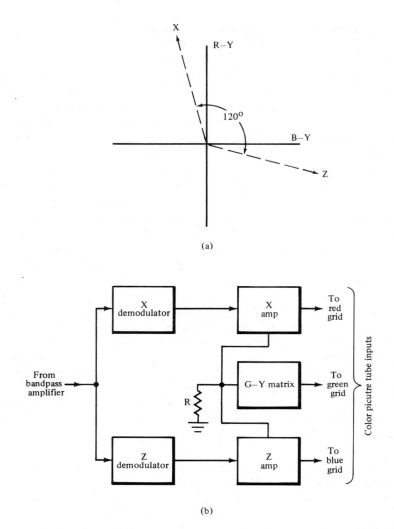

Figure 5–14. Plan of the XZ chroma-demodulation system: **(a)** X and Z demodulation axes; **(b)** Block diagram of demodulator arrangement.

the same manner as in R-Y/B-Y systems. For example, with reference to Fig. 5–15, the output from the X demodulator will be applied to the vertical-input terminal of the vectorscope, and the output from the Z demodulator will be applied to the horizontal-input terminal of the vectorscope. In turn, a normal XZ vectorgram will be displayed with an elliptical contour. Its eccentricity

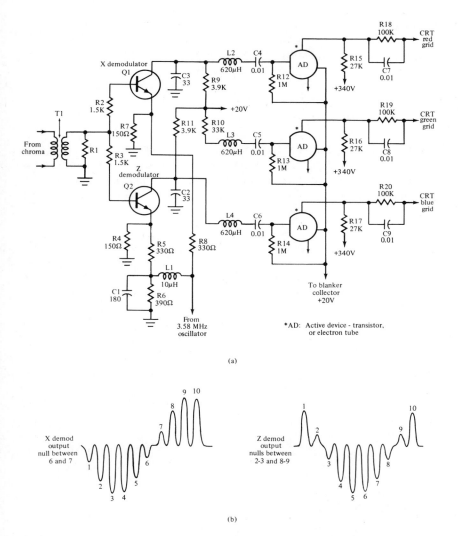

Figure 5–15. Typical XZ chroma-demodulator arrangement: **(a)** Configuration; **(b)** Normal keyed-rainbow output waveforms.

is illustrated by the 120-degree Lissajous pattern in Fig. 5–6. If the vectorgram is circular, or if it has a greater eccentricity than specified, it is indicated that there is a defect in the subcarrier phase-splitting network. With reference to Fig. 5–15, the fault is most likely to be found in C1, R6, or L1. Off values in R4, R5, R7, or R8 can also cause phase-splitting errors. Since L1 cannot be checked with conventional service test equipment, a substitution test should be made if the other trouble possibilities have been eliminated.

5 . 4   TROUBLESHOOTING TUBE-TYPE XZ DEMODULATORS

Next, it is instructive to consider the tube-type XZ chroma-demodulator configuration shown in Fig. 5–16. Its arrangement is basically similar to the solid-state version discussed previously. However, the DC-voltage levels through the circuitry are much

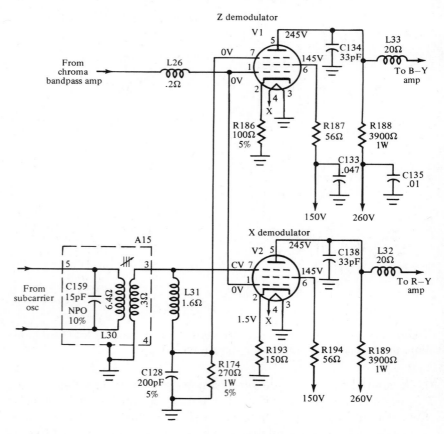

Figure 5–16. A tube-type XZ chroma-demodulator configuration.

higher. As in the case of all tube networks, the first step is to check or replace the tubes in case of circuit malfunction. An oscilloscope can be used to advantage for checking the chroma-signal input to the control grids of the tubes, and for checking the oscillator-signal input at the suppressor grids of the tubes. In other words, an apparent malfunction in the chroma-demodulator section may actually originate in the bandpass-amplifier section or in the subcarrier-oscillator section. Then, if the trouble has been definitely localized to the chroma-demodulator section, DC-voltage and resistance measurements are generally made to pinpoint the defective component. Although it is possible for more than one component defect to occur at the same time in the chroma-demodulator section, this is the exception rather than the rule.

Note that the Z demodulator processes hues in the blue range, and that the X demodulator processes hues in the red range. In turn, when red hues are weak in the image, the X-demodulator section falls under suspicion. To verify or reject the suspicion, the technician makes an oscilloscope check at the outputs of the X and Z demodulators, using a keyed-rainbow signal. The normal amplitudes are 30 volts peak-to-peak for the X demodulator, and 25 volts peak-to-peak for the Z demodulator. Thus, if it is discovered that the output waveform from the X demodulator has an amplitude of, say, 10 volts peak-to-peak, the suspicion is verified. Capacitors in this circuit would then be checked for faults, and resistor values would be measured. Although inductors occasionally become defective, this trouble possibility is checked last.

Sometimes a component defect occurs in the subcarrier input circuitry. As an illustration, C159 in Fig. 5–16 could become leaky or open-circuited. In such a case, the demodulation phases become incorrect, and off-color pictures are reproduced. To check the demodulation phases, a keyed-rainbow input signal is utilized, and oscilloscope waveform checks are made at the outputs of the X and Z demodulators. Typical X and Z demodulation axes (phases) are shown in Fig. 5–17. In turn, the X demodulator normally exhibits peak output on the third and fourth bars (pulses), and null output between the sixth and seventh pulses. Similarly, the Z demodulator normally exhibits peak output on the fifth and sixth bars (pulses), and null output between the eighth and ninth pulses. (See Fig. 5–18.) In case of incorrect demodulation phases, capacitor C128 should also be checked in the example of Fig.

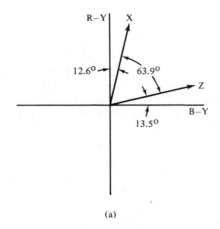

(a)

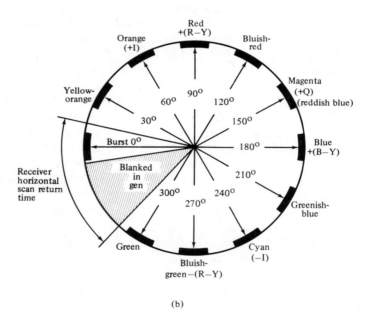

(b)

Figure 5–17. Demodulation phase specification and relations: **(a)** Typical specified X and Z demodulation phases; **(b)** Chroma phase relations in a keyed-rainbow signal.

5–16. Note that the value of R174 is also comparatively critical in the establishment of correct demodulation phases.

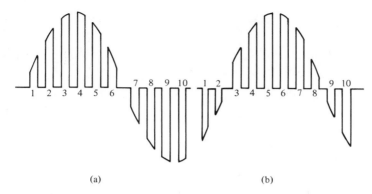

(a)   (b)

Figure 5–18. Normal output waveforms from X and Z demodulators: **(a)** X demodulator output; **(b)** Z demodulator output.

5 . 5   VECTORGRAM DISTORTION CHARACTERISTICS

Troubleshooting is facilitated by an understanding of the causes for vectorgram distortion. Figure 5–19 shows the development of an ideal keyed-rainbow vectorgram. This is the pattern that would be displayed if the keyed-rainbow generator had a perfect output waveform, and if the chroma circuits had very large bandwidth. However, in practice, the input keyed-rainbow waveform is less than ideal, and the chroma channel has comparatively restricted bandwidth. In normal operation, the major restriction is imposed by the chroma-demodulator circuitry, which has a bandwidth of 0.5 MHz. It is helpful to compare the pulse output from a chroma demodulator with a 180-kHz square wave. When such a waveform is passed through a chroma-demodulator circuit, the tops of the waveform become rounded. This distortion is normal, and should not be regarded as a trouble symptom.

Next, it is instructive to consider the causes for nonlinear distortion, as exemplified in Fig. 5–20. This type of distortion causes the vectorgram to be egg-shaped, or off-centered. A check of the demodulator output waveforms will show that at least one of the outputs has unequal peak amplitudes. Thus, Fig. 5–20(a) illustrates a demodulator-output waveform that has less positive-peak amplitude than negative-peak amplitude. Corresponding inequalities are displayed in the positive-peak and negative-peak excursions of the vectorgram. When diode demodulators are em-

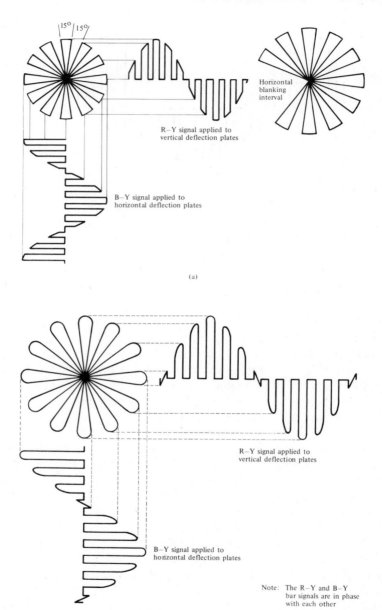

Figure 5–19. Demodulated keyed-rainbow waveforms: **(a)** Ideal waveform; **(b)** Tops are rounded in practice.

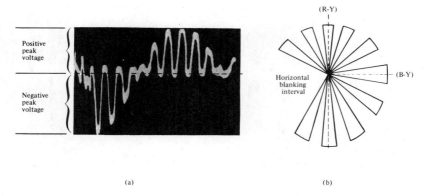

Figure 5–20. Example of nonlinear (off-centered) vectorgram: **(a)** Output waveform from chroma demodulator; **(b)** Nonlinear vectorgram display.

ployed, the front-to-back ratios of the diodes should be checked. Or, if transistor demodulators are utilized, base-emitter bias voltages should be measured. It is also possible that a demodulator transistor has collector-junction leakage; this will show up on a turn-off test. Incorrect bias voltage can cause the transistor to be driven into cutoff or into saturation by a normal chroma-input signal. A leaky collector junction can cause the transistor to be driven into saturation.

Confusion should be avoided between distorted waveforms from the bandpass amplifier, and distorted waveforms from a chroma demodulator. For example, Fig. 5–21 shows an example of distorted waveform output from the bandpass amplifier. Although this input waveform to the chroma demodulator is distorted, the output waveform from the chroma demodulator is unaffected. The reason for this distinction is that the chroma demodulator does not respond on the basis of the peak amplitudes in the waveform of Fig. 5–21. Instead, the chroma demodulator responds to the sine-wave frequency components (fundamental and harmonic frequencies) in the waveform. Therefore, the chroma demodulator normally develops equal positive and negative peaks in its output waveform, in spite of the fact that the input waveform might be distorted as shown in Fig. 5–21.

Next, it is instructive to consider the cause for an "open circle" at the center of a vectorgram, as pictured in Fig. 5–22. Many, although not all, color receivers will show a more or less

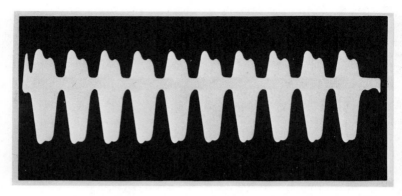

Figure 5–21. Distorted chroma-signal output from the bandpass amplifier.

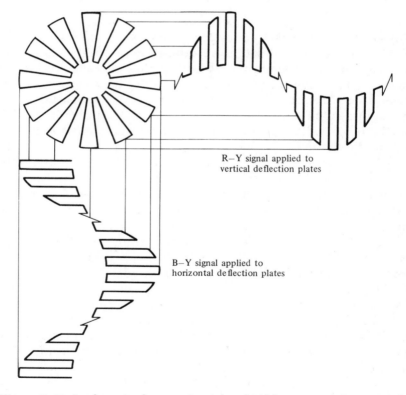

R–Y signal applied to
vertical deflection plates

B–Y signal applied to
horizontal deflection plates

Figure 5–22. Inadequate chroma-circuit bandwidth causes an "open circle"
at the center of the vectorgram.

"open-circle" pattern. This characteristic is caused by base-line curvature in the demodulator output waveforms, as shown in the diagram. As noted previously, this type of base-line curvature results from lack of full high-frequency response in the chroma-demodulator circuits. Unless the opening in the central portion of the vectorgram is considerable, it need not be regarded as a trouble symptom. However, a waveform such as the G-Y waveform shown in Fig. 5–10 has excessive base-line curvature and will produce a very large "open circle" in the center of the vectorgram. In such a case, the G-Y demodulator circuit should be checked out for a defective component.

Note that, although it is general practice to use R-Y/B-Y and XZ vectorgrams, the G-Y output waveform can also be employed to display vectorgrams. As an illustration, Fig. 5–23 shows comparative R-Y/B-Y and G-Y/B-Y vectorgrams. Because the R-Y/B-Y demodulation axes have a 90-degree phase relation, the corresponding vectorgram has a practically circular outline. On the other hand, the G-Y/B-Y demodulation axes have a 56-degree phase relation, approximately. Therefore, the corresponding vectorgram has a prominent elliptical outline. It may be observed that if a G-Y/R-Y vectorgram is displayed, it will have an even more extreme elliptical outline. In other words, the G-Y/R-Y demodulation axes have a 34-degree phase relation, approximately. The ref-

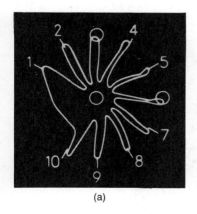

(a)

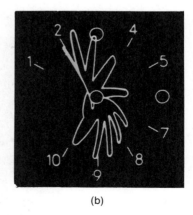

(b)

Figure 5–23. Comparative vectorgrams from three demodulators: **(a)** An R-Y/B-Y vectorgram; **(b)** A G-Y/B-Y vectorgram.

erence ellipses illustrated in Fig. 5–6 show the approximate eccentricities that are displayed by normal G-Y/B-Y and G-Y/R-Y vectorgrams.

To display an elliptical vectorgram properly on the vectorscope screen, the horizontal- and vertical-gain controls of the vectorscope should be adjusted so that the major (long) axis of the ellipse makes a 45-degree angle with the horizontal (and vertical) axes of the screen. When the vertical and horizontal gains are equalized in this manner, the outline of the vectorgram can be properly compared with the reference ellipses illustrated in Fig. 5–6. To repeat a useful point, this comparison can be made to best advantage by temporarily disabling the color-sync section. Then the "petals" in the vectorgram display will rotate rapidly, and the outline of the vectorgram will show up clearly in the pattern.

It is helpful to note that the horizontal-blanking pulse may give the appearance of an "extra petal" in a vectorgram pattern. Many receivers have circuitry that causes the blanking pulse to be displayed in a vectorgram, although some do not. An example is shown in Fig. 5–24; here, the blanking pulse is longer than the ninth and tenth petals, and is almost as wide as both of the petals. The blanking pulse gives the appearance of an "extra petal" in the tenth position. In this situation, the question could be asked whether the blanking pulse might be a petal, or whether the tenth petal might be a blanking pulse. A quick test to distinguish between a petal and a blanking pulse consists in one's turning the color control up and down while watching the vectorgram. Al-

Figure 5–24. Blanking pulse may give the effect of an "extra petal" in a vectorgram.

though the amplitude of the blanking pulse is unaffected by turning the color control, the vectorgram petals will expand or contract in response to the control setting.

REVIEW QUESTIONS

1. What is the function of a chroma demodulator?
2. How are R-Y and B-Y signals separated by a chroma demodulator?
3. Explain how a vectorgram pattern is affected by turning the hue control.
4. Name a component defect that can cause an incorrect demodulation phase angle.
5. Is the resistance value of an inductor related to its inductance value?
6. Why is an oscilloscope helpful in analysis of demodulator trouble symptoms?
7. How are demodulator output waveforms checked for correct phases?
8. Can ATC action simulate chroma-demodulator malfunction?
9. Briefly distinguish between the R-Y/B-Y and the XZ demodulation systems.
10. When malfunction occurs in a tube-type chroma-demodulation circuit, what component or device is checked first?
11. Would a vectorgram have the same waveshape if the chroma demodulators had a bandwidth equal to that of the video amplifier?
12. If a chroma demodulator develops nonlinear output, what component or device would fall under suspicion?
13. Briefly explain the effect of chroma-waveform base-line curvature on vectorgram displays.
14. Does a horizontal-blanking pulse always appear in a vectorgram display?
15. How can a blanking "petal" be quickly distinguished from a vectorgram "petal"?

6

Troubleshooting Chroma and Color Matrices

6.1 GENERAL CONSIDERATIONS

A chroma matrix combines two or more signal sources to form another signal. As an illustration, a G-Y matrix combines R-Y and B-Y signals in suitable proportions to form a G-Y signal. A color matrix combines the Y signal with one or more chroma signals to form a color signal. For example, a color picture tube may be operated as an RGB matrix, to form red, blue, and green color signals from the Y signal, R-Y signal, B-Y signal, and G-Y signal. The basic principle of a G-Y matrix is shown in Fig. 6–1. In this example, a green signal is being processed. Accordingly, there is a 0.59 Y-signal output, a 0.41 G-Y matrix output, a 0.59 R-Y signal output, and a 0.59 B-Y signal output. Thus, the G-Y matrix develops an output signal that adds with the Y signal to give full output from the green gun. At the same time, the R-Y and B-Y signals are cancelled by the Y signal, so there is zero output from the red gun and the blue gun. In case of component defects, there may be less than full output from the green gun, and there may be spurious outputs from the red and blue guns.

To check the output from the G-Y matrix, and to compare it with the R-Y and B-Y signals, a scope is used, as depicted in Fig. 6–2. A keyed-rainbow generator is generally used to energize the receiver. At any instant, the G-Y matrix combines 0.51 of the

136

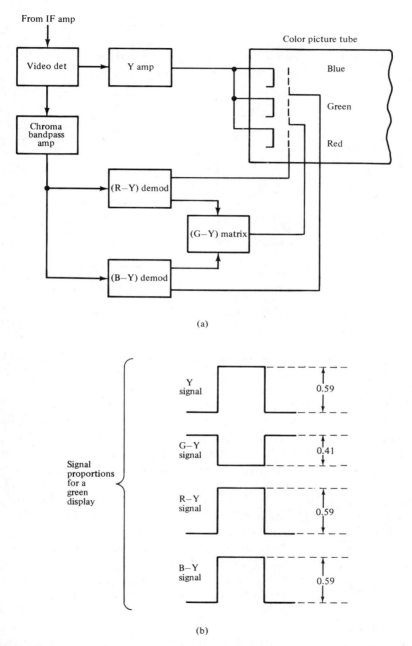

Figure 6–1. Basic principle of a G-Y matrix: **(a)** Block diagram; **(b)** Demodulator and matrix output waveforms for a green-bar display.

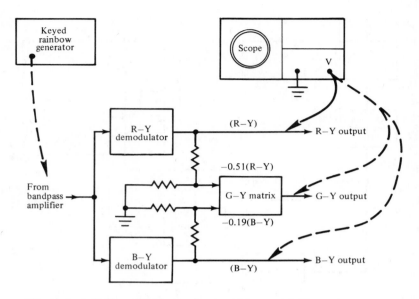

Figure 6–2. R-Y and B-Y signal proportions at G-Y matrix input.

-(R-Y) signal with 0.19 of the -(B-Y) signal to form the corresponding G-Y signal. First, the technician notes the amplitude of the G-Y signal, as shown in Fig. 6–3. In other words, the G-Y signal (at the picture tube) normally has an amplitude that is 0.57 of the R-Y signal amplitude, and 0.32 of the B-Y signal amplitude (when a keyed-rainbow signal is being processed). If the G-Y signal amplitude is substantially incorrect, the associated circuitry is checked for a defective component. Note in passing that all color receivers do not employ the exact chroma-signal proportions indicated in Fig. 6–3. This depends to some extent upon the phosphors used in the color picture tube. However, the receiver service data will usually specify chroma-signal peak-to-peak voltages.

A basic G-Y matrix configuration is shown in Fig. 6–4. Mixing action occurs in R1, R2, R3, and the base circuit of Q. Amplification is provided by the transistor. Gain is determined chiefly by the value of R4, although the value of R5 also has some effect on gain. If the transistor develops collector leakage, the gain will be reduced. This condition can be checked by making a turn-off test of Q. A matrix stage normally operates in class A, as determined by the base-bias voltage on the transistor. If a scope test shows

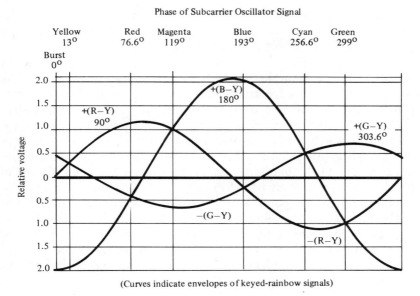

(Curves indicate envelopes of keyed-rainbow signals)

Figure 6–3. Typical amplitude relations of the R-Y, B-Y, and G-Y signals at the grids of the color picture tube. *(Redrawn by permission of RCA)*

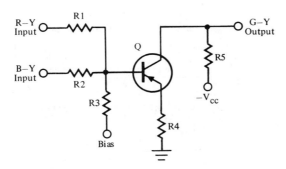

Figure 6–4. Basic G-Y matrix configuration.

that the matrix output is nonlinear, with the positive-peak amplitude greater than the negative-peak amplitude, or vice versa, the bias voltage should be measured. If the G-Y waveform crossovers are incorrect, although the R-Y and B-Y waveform crossovers are correct, the values of the matrix resistors R1, R2, and R3 should be checked. Some matrix configurations employ capacitors, in addition to resistors and transistors. In such a case, the capacitors are prime suspects in associated trouble symptoms.

Typical chroma output waveform displays are illustrated in Fig. 6–5. In this example, the vertical-gain control of the scope has been adjusted so that each waveform appears with the same amplitude. A normally operating R-Y/B-Y/G-Y demodulator-matrix system will display similar waveforms with the crossovers indicated, and with equal positive- and negative-peak voltages. Note the blanking pulses preceding the first pulse and following the tenth pulse. These blanking pulses normally have a considerably greater amplitude than the chroma pulses. In some receivers, the blanking pulses are comparatively wide, and may tend to obscure the first pulse, for example. In this situation, it is helpful to change the setting of the horizontal-hold control. The resulting change in timing of the blanking pulse shifts its position with respect to the chroma pulses. Thus, a partially concealed pulse can be shifted into view.

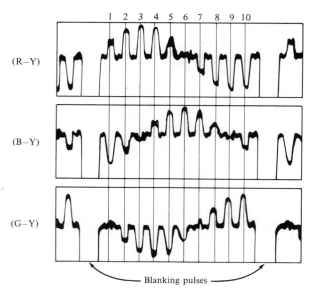

Figure 6–5. Output waveforms from chroma demodulators with receiver energized by a keyed-rainbow color-bar signal. *(Redrawn by permission of RCA)*

6.2 MATRIXING IN THE XZ SYSTEM

An XZ chroma-demodulation system may utilize various demodulator phases. Previous mention was made of an XZ system that
employs a 120-degree phase difference between the X and Z axes.
Another example is shown in Fig. 6–6. This arrangement utilizes a
67-degree phase difference between the X and Z axes, with sub-
sequent matrixing action that develops R-Y, B-Y, and G-Y outputs.
Normal matrix action requires correct chroma-demodulator phase
angles. Transistors Q1 and Q2 operate in combination with matrix
resistor R1 to provide the matrixing function. A secondary effect
on matrixing action is provided by R2, R3, and R4. In case of
trouble symptoms, the chroma-demodulator phase angles should

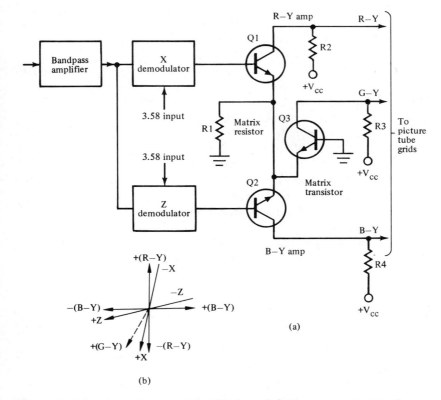

Figure 6–6. Basic matrix circuit for XZ demodulation system: **(a)** Configuration; **(b)** Chroma demodulation axes.

be checked first. This check is made on the basis of the specified crossovers for the X and Z demodulator output waveforms. If demodulator action is normal, the technician proceeds to analyze the matrix circuit action.

Direct current voltages are measured first in the matrix section. Erroneous values may or may not be conclusive, because of the direct coupling that is used, with resulting circuit interactions. Therefore, follow-up tests are often required. With reference to Fig. 6–6, turn-off tests of transistors Q1, Q2, and Q3 will show whether a transistor may have collector-junction leakage. As a practical note, the output lead from a matrix transistor is likely to have a return path to ground from the picture-tube input circuit. For example, the G-Y output lead from Q3 probably has a resistive path to ground at the green-grid terminal of the picture tube. Therefore, the technician makes a razor slit in the G-Y output lead between R3 and the picture-tube input circuit. In turn, after the turn-off test of Q3 has been completed, the slit is repaired with a small drop of solder.

If the transistors respond normally to turn-off tests, the technician concludes that there is an off-value resistor in the network. With reference to Fig. 6–6, the value of R1 can be measured in circuit with a lo-pwr ohmmeter. It may be possible to measure the values of R2, R3, and R4 in circuit also; however, this possibility depends upon the picture-tube input circuitry. When there is a shunt path to ground, the technician may razor-slit the associated PC conductor, as explained previously, so that the resistor value can be measured out of circuit. Or the technician may prefer to unsolder one end of the resistor from the circuit board for this test. Note in passing that puzzling trouble symptoms are occasionally caused by cold-soldered connections, cracked circuit boards, or foreign substances that cause leakage between circuit-board conductors.

6.3   MATRIXING IN RGB SYSTEMS

An RGB matrixing system is a color-matrix arrangement, instead of a chroma-matrix arrangement. To make this distinction clear, refer to Fig. 6–1. This is an example of a chroma-matrix configuration, followed by RGB (color) matrixing in the color picture tube. The essential point is that the Y signal is combined with the chroma signals in the picture tube. On the other hand, various color receivers use the color picture tube only as a display device, with the RGB matrix action taking place in a separate RGB matrix. Still other receivers employ RGB matrix action in the chroma-demodulator circuitry. In this case, the circuits operate as combination demodulator-matrices. It is important for the technician to recognize the type of matrixing action that is utilized. Otherwise, it becomes difficult to analyze trouble symptoms effectively.

A block diagram for a color-matrix system with an RGB matrix apart from the color picture tube is shown in Fig. 6–7. Note that the cathodes of the picture tube are driven by the color signals in this arrangement. Only DC bias voltage is applied to the grids of the color picture tube. A practical point concerning vectorscope checks of cathode-driven circuitry is seen in Fig. 6–8. Observe that the horizontal-blanking interval appears in the upper right-hand portion of the vectorgram, compared with blanking in the lower left-hand portion of the pattern when the grids are driven in the color picture tube.

A somewhat more detailed block diagram for the separate RGB matrix system is seen in Fig. 6–9. R-Y, B-Y, and G-Y chroma signals are demodulated and then combined with the Y signal in

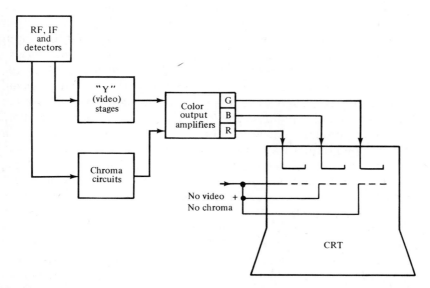

Figure 6–7. Block diagram for an external RGB color-matrix arrangement.

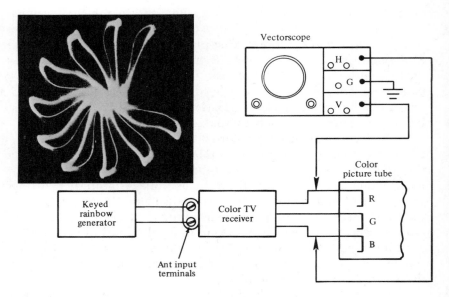

Figure 6–8. Typical vectorgram displayed by a cathode-driven picture-tube arrangement.

the color-video output stages, which function as color matrices. Individual drive controls are provided for the color-video output stages, so that the amplitudes of the color-picture-tube input waveforms can be adjusted as specified in the receiver service data. Adjustable DC bias voltages are applied to the grids of the color picture tube. Troubleshooting of this RGB matrixing system is based on the same principles as for the conventional system that uses the color picture tube for final matrixing. However, there are the additional color-video output stages to be checked out. In normal operation, a waveform check will show that the output signal is a replica of the input signal, but at a higher amplitude. When trouble symptoms are encountered, the technician proceeds with DC-voltage measurements, transistor cutoff tests, and resistance measurements.

Figure 6–10 shows a configuration for a typical R matrix arrangement. Observe that the R-Y signal is applied to the base of the color-video transistor, and the Y signal is applied to the emitter of the transistor. In turn, color matrixing occurs in the base-emitter circuit. When malfunction occurs, note that DC-voltage measurements may sometimes be inconclusive because of the ex-

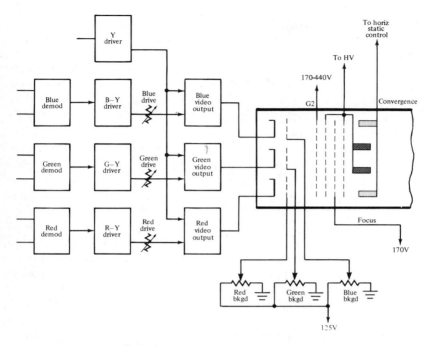

Figure 6–9. Block diagram for a separate RGB matrix system.

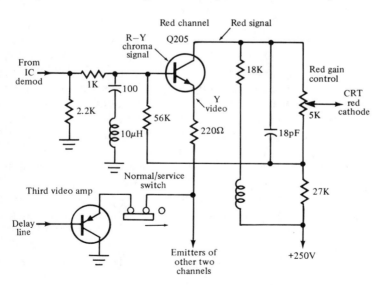

Figure 6–10. An R matrix configuration.

tensive direct-coupled circuitry that is involved. Therefore, it is often helpful to make a turn-off test of the transistor. In this example, a conclusive test requires that the shunt path through the red gain control be temporarily opened. In other words, a razor slit may be cut in the PC conductor between the 18K resistor and the 18pF capacitor. Then a turn-off test may be made in the usual manner. After the test is completed, the razor slit is repaired with a small drop of solder.

As a practical troubleshooting note, two fixed capacitors are included in the configuration of Fig. 6–10. If the 100pF capacitor becomes "open," there will be excessive subcarrier interference in the picture. Or, if this capacitor becomes "shorted" or leaky, the base-bias voltage on Q205 will be subnormal. If the 18-pF capacitor becomes short-circuited, the red gain control will be inoperative. Or, if this capacitor becomes "open," the frequency response of the stage will become subnormal. Inductors seldom cause trouble symptoms. However, peaking coils and traps can be mechanically damaged if accidentally struck by a tool or overheated during a soldering operation. If all of the other components check out normally, the technician will conclude that an inductor is defective, and will make substitution tests. Note that if an inductor appears to be "open" when a continuity test is made, the connections of the fine wire to the pigtail leads should be investigated before it is assumed that the trouble is internal.

Next, consider the simultaneous RGB demodulation and matrixing arrangement shown in Fig. 6–11. Note that the Y signal is introduced into the color demodulators with the R-Y, B-Y, and G-Y signals. Thus, each demodulator processes the chroma signal combined with the Y signal. Accordingly, the demodulation phases are along the primary color axes, instead of the chroma axes. Note that the diodes in the green-demodulator circuit are polarized in a sequence opposite to that of the diodes in the red-demodulator and blue-demodulator circuits. This circuit means is employed to automatically cancel out spurious demodulation products called "blips." Because of the diode polarity reversal in the green-demodulator circuit, the subcarrier is necessarily injected in the magenta phase, which is 180 degrees from the green phase, as seen in Fig. 6–12.

Troubleshooting a simultaneous G-Y demodulation and matrixing configuration involves the same general procedures as previously discussed for the separate RGB matrix system. Thus, wave-

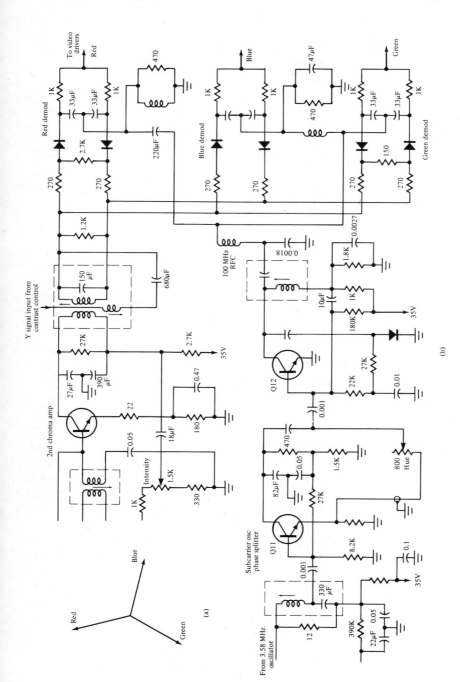

Figure 6-11. Simultaneous RGB demodulation and matrixing arrangement: **(a)** Demodulator phases; **(b)** Typical configuration.

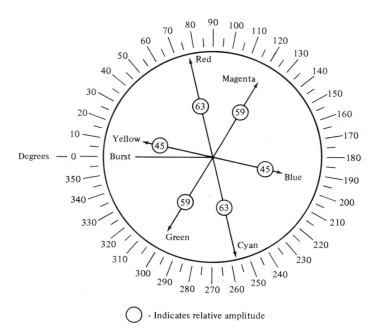

○ - Indicates relative amplitude

Figure 6–12. Phase angles of the primary and complementary colors.

form checks are made to determine whether an input signal may be weak, distorted, or absent, and to observe whether the output waveforms have correct amplitudes and crossovers as specified in the receiver service data. Note that during reception of a black-and-white program, the Y signal feeds through the color demodulator circuits, and no demodulation occurs. In other words, the input Y waveform has the same shape as the output Y waveform. On the other hand, during reception of a color program, the chroma component and the subcarrier-oscillator signal cause the circuits to function as demodulators.

Another configuration for simultaneous RGB demodulation and matrixing is seen in Fig. 6–13. Chroma signal is applied to the primary of T1, and Y signal is applied at the center tap on the secondary. Subcarrier voltage is coupled via C1 and C2 to the cathode of D1 and to the anode of D2. During black-and-white program reception, the signal feeds through D1 and D2 without any demodulating action. On the other hand, during color program reception, there is a chroma signal input to the primary of T1. In turn, the chroma-signal phase combines with the subcarrier phase

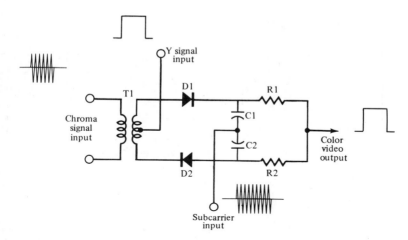

Figure 6–13. Another common configuration for simultaneous RGB demodulation and matrixing.

at diodes D1 and D2, resulting in unequal conduction paths to R1 and R2. Thus, chroma demodulation takes place, and the demodulated chroma signal is superimposed on (matrixed with) the Y signal to form the color video output waveform. Troubleshooting this arrangement is basically the same as was explained for the configuration of Fig. 6–11.

It is helpful for the troubleshooter to be familiar with the demodulator-matrix circuit-board configuration shown in Fig. 6–14. This is an integrated-circuit arrangement that is frequently encountered. Note that the 3.58-MHz subcarrier voltage from terminal 2 on the circuit board is coupled to the phase-shift network comprising L1, R1, C1, and the tint (hue) control (not shown). The phase of the subcarrier at the junction of L1 can be shifted about 70 degrees by the viewer, in normal operation. The subcarrier is coupled via C2 to terminal 4 of the demodulator integrated circuit IC301. Next, the phase of the subcarrier is shifted 103 degrees by L2, R2, and C4. This subcarrier phase is coupled by C3 to terminal 5 of IC1. Note that the chroma signal is available at pin 1 on the circuit board and is coupled via C5 to terminal 3 on IC1.

As Fig. 6–15 shows, this integrated circuit contains two doubly balanced synchronous demodulators. In other words, the IC configuration is balanced with respect to both of the input signals — the subcarrier signal and the chroma signal. The subcarrier cancels itself out in the demodulators and does not feed through

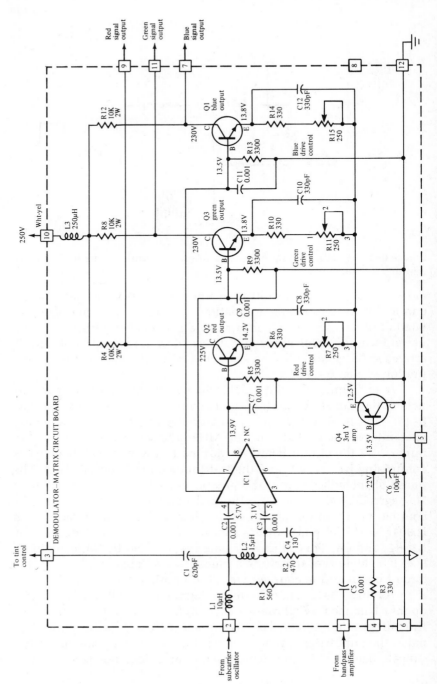

Figure 6-14 Configuration of a demodulator-matrix circuit board.

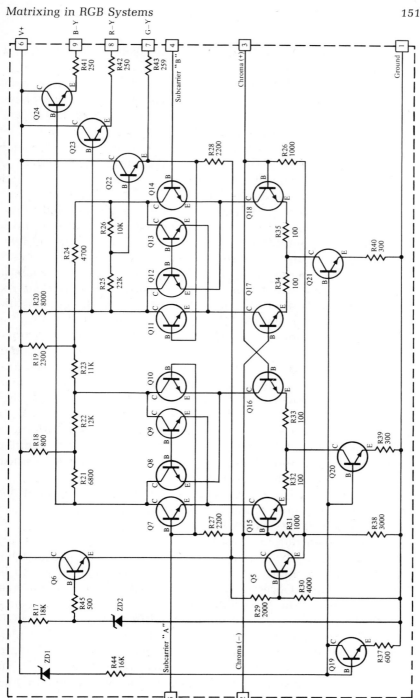

Figure 6–15. IC pair of doubly balanced synchronous detectors.

into the R-Y, B-Y, and G-Y output circuits. This action eliminates the need for including low-pass filters in the output leads. Note in Fig. 6–15 that the chroma signal at terminal 3 of the IC is connected to the parallel inputs of Q16 and Q18. Opposite-phase chroma signals from the differential amplifier, Q15 and Q16, are fed to the transistor switches, Q7, Q8, Q9, and Q10. Opposite-phased chroma signals from the differential amplifier, Q17 and Q18, are fed to the transistor switches Q11, Q12, Q13, and Q14. These switches are controlled by two signals: the subcarrier signal at pin 4, which is also applied to the bases of Q11 and Q14; and the second subcarrier signal at pin 5, which is also applied to the bases of Q7 and Q10. Note that the chroma signal currents from each differential amplifier will flow into one of the two outputs, depending on the instantaneous states of the individual switches.

Next, we observe in Fig. 6–15 that the outputs from each of the two synchronous detectors are fed to a resistive chroma matrix comprising resistors R18 through R26, where the three color-difference signals are formed. A regulated 6-volt supply for the switching transistors is provided by Zener diode ZD302 and transistor Q6. This supply voltage flows through a voltage divider and is applied to the base of Q5 to establish a 3-volt supply for the differential amplifiers. Transistors Q20 and Q21, which are controlled by the diode-connected transistor Q19, are regulated current sources for the differential amplifiers. Finally, the R-Y, B-Y, and G-Y output voltages from the chroma matrix are fed through the emitter-follower resistors Q22, Q23, and Q24. Each emitter lead contains a 250-ohm resistor, which protects the IC against damage from accidental short circuits.

With reference to Fig. 6–14, note that DC bias voltage and the color signal for the red output amplifier, Q2, are developed across R5, which is bypassed by C7. Similarly, DC bias and the color signal for the green output amplifier, Q3, are developed across R9, which is bypassed by C9. Also, DC bias and the color signal for the blue output amplifier, Q1, are developed across R13, which is bypassed by C11. Observe that the Y signal is injected into the emitter circuits of the color output amplifiers from the third Y amplifier, Q4. In turn, the color output amplifiers operate as an RGB matrix. The red, green, and blue signal outputs are applied respectively to the red, green, and blue cathodes of the color picture tube. Note that postdemodulator RGB matrixing is employed in Fig. 6–14, whereas predemodulator matrixing is utilized in Fig.

6–11. Owing to this distinction, demodulation takes place on the R-Y, B-Y, and G-Y axes in the configuration of Fig. 6–14, whereas demodulation proceeds on the R, G, and B axes in the example of Fig. 6–11.

Troubleshooting in the IC arrangement of Fig. 6–14 starts with analysis of the trouble symptoms. As an illustration, if the green hue is absent from the color image, the technician directs his attention to the circuitry associated with Q3 and IC1. DC voltages are usually measured first, and compared with the specified values. If Q3 comes under suspicion, a turn-off test may be made, as explained previously. This test may reveal, for example, that Q2 is in good condition, and that the trouble is being caused by leakage or a short circuit in C7. Again, it may be determined that subnormal output from Q2 is being caused by an "open" in C8. In case of doubt when one is evaluating DC voltage measurements, it is often helpful to follow up with in-circuit resistance measurements, using a hi-lo ohmmeter.

6 . 4   SERVICING TUBE-TYPE MATRIX CIRCUITRY

Consider next the typical tube-type G-Y matrix circuit shown in Fig. 6–16. Outputs from the R-Y and B-Y demodulators are suitably proportioned through voltage dividers and combined to form the G-Y signal. Note that the test receptacle in the R-Y demodulator circuit is used for DC-voltage checks; normally, a positive voltage will be measured that is proportional to the amplitude of the incoming chroma signal. Tubes are replaced first, in case of faulty operation. Then, if the trouble symptom persists, the waveforms at the R-Y, B-Y, and G-Y outputs should be checked. Figure 6–17 shows normal waveforms for a keyed-rainbow generator input signal. Observe that the G-Y waveform nulls on the first bar (pulse), and on the seventh bar.

Incorrect G-Y nulls throw suspicion upon the grid circuit of the G-Y matrix (amplifier) tube. In other words, the technician would check C221, R222, R223, and R224. It is assumed that the blue gain control, R211, is set correctly, so that the R-Y and B-Y output waveforms have normal relative amplitudes, as specified in the receiver service data. Observe that the relative amplitudes of the R-Y, B-Y and G-Y waveforms are not the same in all color receivers. These differences depend upon the type of color picture

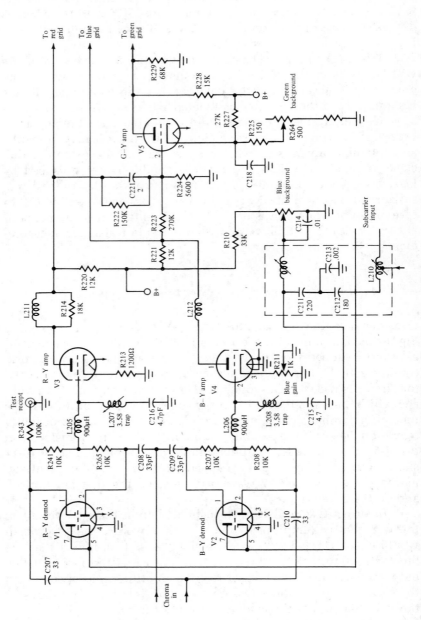

Figure 6–16. A typical tube-type G-Y matrix configuration.

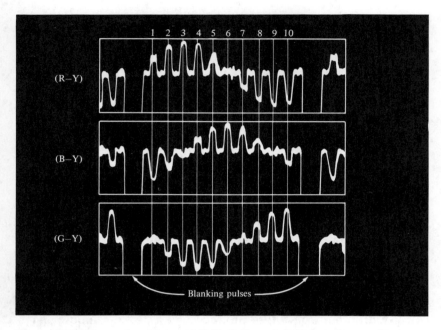

Figure 6–17. Example of normal R-Y, B-Y, and G-Y waveforms. *(Redrawn by permission of Sencore, Inc.)*

tube that is employed. In other words, different varieties of phosphors have different comparative light outputs. In Fig. 6–17, the vertical-gain control of the oscilloscope has been set to display the waveforms at uniform amplitudes, since this is basically a relative phase illustration.

6 . 5   NOTES ON NTSC COLOR-BAR GENERATORS

Although NTSC color-bar generators are used chiefly in laboratories and color-TV broadcast stations, the service technician will also find good use for this type of generator upon occasion. As an illustration, RGB matrix action is checked to best advantage with an NTSC signal, because it contains a Y component. By way of comparison, a keyed-rainbow signal has no Y component. Figure 6–18 shows the appearance of an NTSC color-bar generator. Its ba-

Figure 6–18. Appearance of an NTSC color-bar generator. *(Courtesy of Tektronix)*

sic output waveforms are illustrated in Fig. 6–19. These Y, chroma, and complete color-signal outputs are individually available. Accordingly, the matrixing action can be tested for response to a Y signal alone, to a chroma signal alone, and to the complete color signal. In addition, an NTSC generator provides R-Y and B-Y bar signals, as depicted in Fig. 6–20. These are very useful in checking R-Y and B-Y demodulator action.

Vectorgrams produced by an NTSC signal are characteristically different from those produced by a keyed-rainbow generator. Figure 6–21 shows a typical NTSC vectorscope graticule, and how the pattern consists basically of bright dots. In other words, the electron beam in the CRT jumps rapidly from one color-bar value to the next, and rests in the position of the corresponding color for the duration of each color bar. White and black signals produce a bright spot in the center of the NTSC vectorgram. Note that the box size indicated in Fig. 6–21 shows the permissible tolerance on the color signal for a phase error of ±5 degrees, and for an amplitude error of ±8 percent. These are the tolerances that are usually observed in color-TV broadcast operation.

It should be noted that the color-bar sequence in an NTSC signal is arbitrary. In other words, the sequence illustrated in Fig. 6–19 is not utilized by all manufacturers. Another sequence of the primary and complementary color bars is shown in Fig. 6–22. Still another sequence was shown in the preceding chapter. Most NTSC color-bar generators provide supplementary test signals such as G-Y and G-Y /90-degree outputs, I and Q signal outputs. Some generators also provide a simultaneous color-bar pattern that includes the primary and complementary colors with I and Q bars. The latter are utilized only by technicians in color-TV broadcast stations.

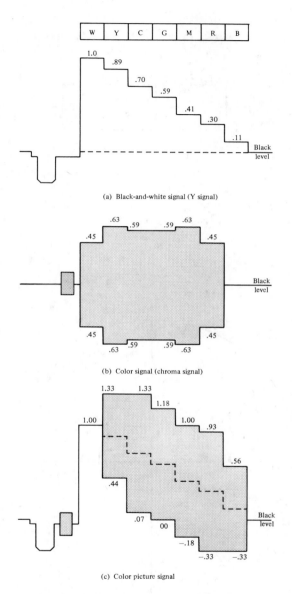

(a) Black-and-white signal (Y signal)

(b) Color signal (chroma signal)

(c) Color picture signal

Figure 6–19. Output waveforms from an NTSC color-bar generator.

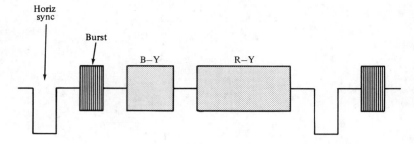

Figure 6–20. R-Y and B-Y chroma-bar output from an NTSC generator.

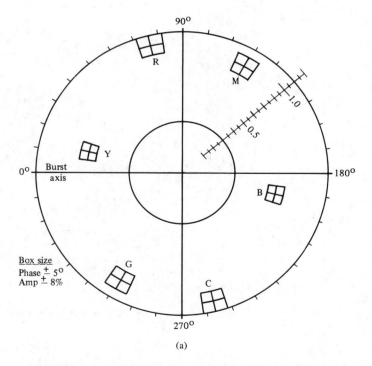

(a)

Figure 6–21. NTSC vectorgram characteristics: **(a)** Vectorscope graticule.

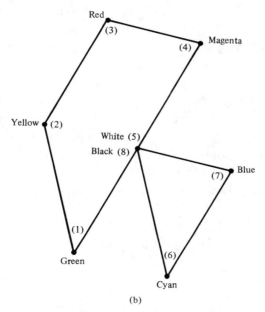

Figure 6–21. (b) Development of NTSC vectorgram.

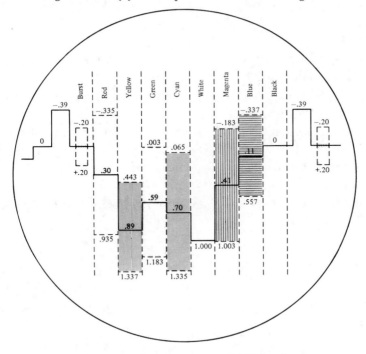

Figure 6–22. Another sequence of NTSC color bars.

REVIEW QUESTIONS

1. What is the function of a chroma matrix?
2. How is an oscilloscope employed to analyze chroma-matrix action?
3. Can a G-Y signal be matrixed from X and Z signals?
4. Explain how a component can be disconnected from a circuit board without unsoldering its terminals.
5. Briefly discuss how a cold-soldered connection can simulate a component or device defect.
6. Define an RGB matrix system.
7. Discuss the operation of a color picture tube as an RGB matrix.
8. Why is an oscilloscope often helpful in analyzing matrix malfunctions?
9. Distinguish between pre-demodulator matrixing and post-demodulator matrixing.
10. Are matrix output signal proportions the same in all receivers?
11. Describe the utility of an NTSC-type color-bar generator in troubleshooting RGB matrix circuitry.
12. How does an NTSC-signal vectorgram differ from a keyed-rainbow vectorgram?
13. Is the color-bar sequence always the same in a keyed-rainbow pattern?
14. Is the color-bar sequence always the same in an NTSC pattern?
15. Distinguish between G-Y and G-Y /90-degree signal outputs.

Systematic Color-circuit Troubleshooting

7 . 1   GENERAL CONSIDERATIONS

Each technician develops his own systematic troubleshooting procedure. Some of these procedures are more efficient than others. If a shop is well equipped with test and measuring instruments, the technician has more options than in a poorly equipped shop. Some technicians tend to avoid the use of the more sophisticated instruments, such as the oscilloscope. This is an unfortunate situation, because an oscilloscope is a very informative instrument when its application is understood. Technicians who seldom or never use an oscilloscope usually employ signal-substitution equipment, such as a television analyst. This instrument can also be very informative in preliminary analysis of trouble symptoms. The following topic illustrates the systematic use of a television analyst in color-circuit troubleshooting.

7 . 2   NO COLOR REPRODUCTION; BLACK-AND-WHITE RECEPTION NORMAL

This trouble symptom can be caused by a failure in the color-IF amplifiers, or in the color-killer section. Localization tests proceed as follows:

1. With a television analyst as a signal source, inject a high-level color signal at the input of the color demodulators. It is improbable that all of the demodulators would fail simultaneously. Therefore, the technician assumes that this trouble symptom is caused by a defective section other than the demodulator section. If color is displayed on the picture-tube screen when the color signal is injected at the input of the color demodulators (Fig. 7–1), it is established that the subcarrier oscillator is operating. Figure 7–2 shows a skeleton circuit diagram to denote the test points that are involved. Since color is displayed on the screen, the technician proceeds to step 2.

2. Inject a high-level color signal at the output of the second color-IF amplifier. If no color is displayed on the picture-tube screen, check the coupling between the second color-IF amplifier and the color demodulators. Signal-injection tests are continued to isolate the defect to as small an area as possible. On the other hand, if color is displayed on the screen, the technician proceeds to step 3.

3. Next, check the gain of the second color-IF amplifier by moving the signal-injection point to the input of the stage and observing the difference in color intensity on the screen. If the amplifier stage is operating properly, the intensity of the color display should increase greatly when the signal-injection point is changed to the input of the stage. The relative amount of gain can be checked more closely by reducing the setting of the color-intensity control until the color display is barely visible on the screen with the color signal injected at the output of the stage, and then by moving the injection point to the input of the stage, and again reducing the setting of the color-intensity control setting until the color display is again barely perceptible.

   If the stage gain is approximately normal, the technician proceeds with step 4. On the other hand, if the gain is low, or if no color display is obtained with the signal injected at the input of the stage, the trouble will possibly be found in the second color-IF amplifier circuit, or in the color-killer circuit. In this situation, the technician proceeds to step 6 for isolation procedure.

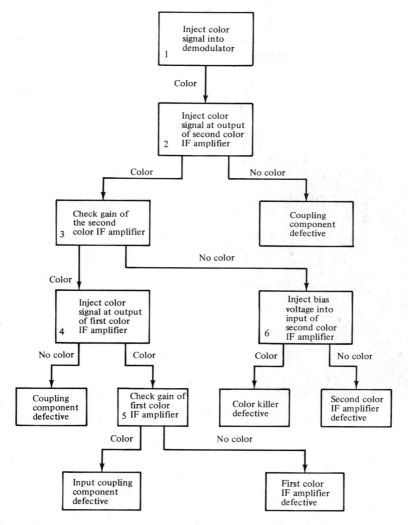

Figure 7–1. Troubleshooting procedure for a no-color symptom. *(Redrawn by permission of B & K Mfg. Co. Division of Dynascan Corporation)*

4. A medium-amplitude color signal is injected next at the output of the first color-IF amplifier. If no color is displayed on the screen, the coupling components between the first color-IF amplifier and the second color-IF ampli-

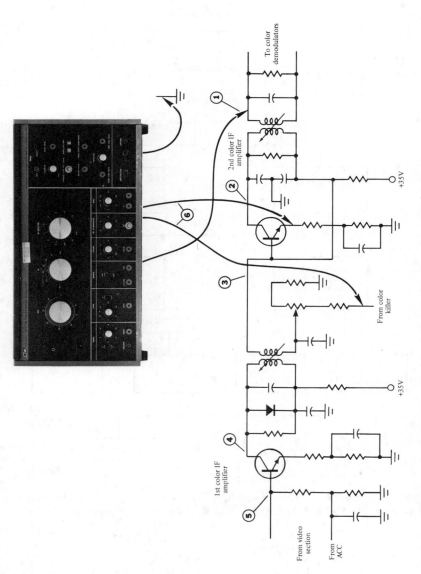

Figure 7-2. Color-IF and color-killer test points for signal-injection tests. (*Courtesy of B & K Mfg. Co. Division of Dynascan Corporation*)

fier are at fault. On the other hand, if a color display is obtained, the technician refers back to step 3 for procedure.

5. The gain of the first color-IF amplifier is now checked, according to procedure in step 3. If the gain is normal, the coupling components between the video-amplifier section and the first color-IF amplifier should be checked. On the other hand, if the gain is low, or if no color display is obtained, it is indicated that the first color-IF stage is inoperative. In turn, the technician proceeds to close in on the component that has disabled the stage.

6. This step serves to isolate a fault to the second color-IF amplifier or to the color killer. Note that the DC output of the color killer normally keeps the amplifier cut off until a color signal is received. If the color killer is defective, the amplifier will remain cut off although a color signal is present. A VHF signal is injected at the antenna terminals of the receiver, and the color control is turned up. A bias power supply is connected to the input of the second color-IF amplifier, and the bias control is adjusted to ensure that the amplifier is not cut off. If the color display is restored, it is indicated that the fault is in the color-killer stage. However, if no color is displayed, the defect will be found in the second color-IF amplifier.

7 . 3   ONE COLOR ABSENT IN SCREEN PATTERN

This trouble symptom affects one of the primary colors; thus, the red bar number 3 will be colorless, or the blue bar number 6, or the green bar number 10. In other words, when the hue control is adjusted to display two of the primary colors correctly, the third primary color is absent. The remaining color bars generally have incorrect hues because of incorrect primary blends. This trouble symptom could be caused by an inoperative chroma demodulator, defective color amplifier stage, or a picture-tube fault. Troubleshooting procedures are essentially the same for the red, blue, or green color circuits. Accordingly, the following procedure discusses the red color channel only. Refer to Figs. 7–3 and 7–4.

1. Disconnect the antenna from the receiver and disable the

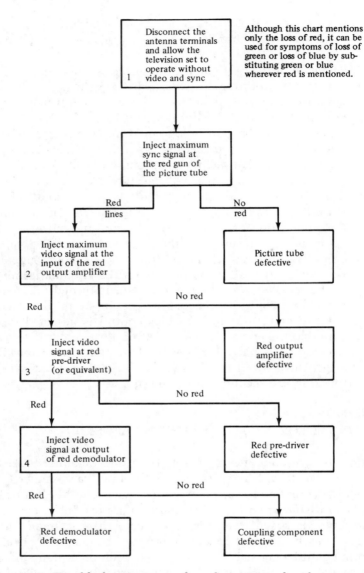

Figure 7–3. Troubleshooting procedure for a one-color-absent symptom. *(Redrawn by permission of B & K Mfg. Co. Division of Dynascan Corporation)*

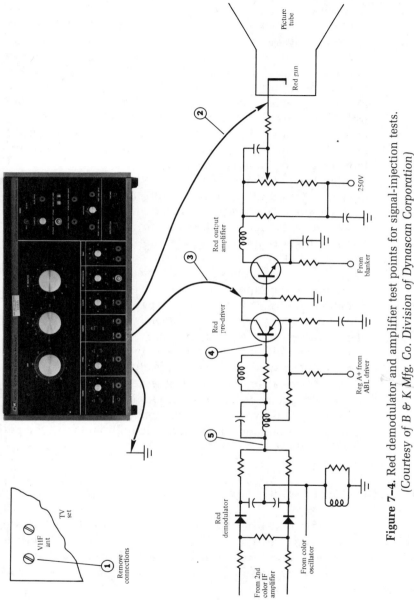

Figure 7-4. Red demodulator and amplifier test points for signal-injection tests. (Courtesy of B & K Mfg. Co. Division of Dynascan Corporation)

first video amplifier stage. For example, the base and emitter terminals of the transistor can be temporarily short-circuited.

2. Inject a high-level sync signal with positive polarity at the red gun of the picture tube. In normal operation, several red diagonal bars will be displayed. The bar pattern can be "free-wheeled" by adjusting the horizontal-hold control. If red bars are visible, it is indicated that the red gun is operative, and the technician proceeds to step 3. On the other hand, if red bars are not displayed, it is indicated that the picture tube is defective.

3. A high-level video signal is now injected at the input of the red output amplifier. Both positive-polarity and negative-polarity signals should be utilized. If red bars are displayed in either test, proceed to step 4. On the other hand, if there is no red display, the technician injects the sync signal at various points in the output circuitry of this stage to localize the fault insofar as possible.

4. If the receiver employs a red predriver or similar stage, as depicted in Fig. 7–4, a video signal is injected into this stage. If a red pattern is displayed, the technician proceeds to step 5. However, if there is no red display obtained, it is indicated that the predriver stage is defective.

5. Next, a video signal is injected at the output of the red demodulator. If a red video display is obtained, the trouble will be found in the demodulator circuit. On the other hand, if a red video display is not obtained, the technician will check the coupling components between the demodulator and the red amplifier stage.

7 . 4   NO COLOR SYNC

Loss of color sync is caused by lack of locking in the 3.58-MHz subcarrier and APC section. It can result from a faulty burst-amplifier stage, a phase-detector defect, reactance-control malfunction, or off-frequency oscillator operation. Refer to Fig. 7–5 and proceed as follows:

1. First, inject bias voltage at the input of the reactance stage and vary the bias control setting. If the reactance stage is

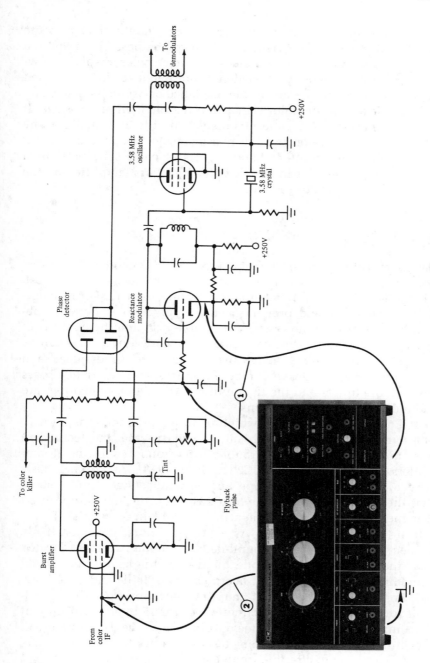

Figure 7-5. Color-sync circuit test points for signal-injection tests. (*Courtesy of B & K Mfg. Co. Division of Dynascan Corporation*)

operating, the hue of the color bars should change as the bias control is varied, and it should be possible to apply a bias voltage that produces a correct color display. However, if the normal reaction is not obtained, it is indicated that there is a fault in the reactance or oscillator stage.

2. Next, inject a color signal into the burst-amplifier stage. If a normal color pattern is displayed, it is indicated that the burst-amplifier and the phase-detector sections are workable. On the other hand, if loss of color sync persists, a defect will be found in either the burst-amplifier or the phase-detector section.

7.5 FLOW-CHART ANALYSIS

Sectionalization and localization procedures are often facilitated by considering the signal flow chart for the receiver under test. Figure 7–6 shows a flow chart for a widely used design of color-TV receiver. As noted previously, the basic troubleshooting problem is to determine the cause of a trouble symptom by analysis of receiver response and test data. In a flow chart, four signals can be recognized: the chroma signal, the Y signal, the sync signal, and the sound signal. Observe that all four signals normally pass through the tuners and the IF amplifiers, thereupon branching into the individual receiver sections.

Consider a situation in which picture reproduction is normal, but sound reproduction is absent. In such a case, the technician does not expect to find the trouble in the tuners or in the IF amplifiers. In conclusion, test IC201, Q801, Q201, or Q803. If tests with a wide-band oscilloscope and low-capacitance probe show that there is no signal voltage to IC201, then the first supposition was incorrect, and the trouble will be found in either the IF section or in the tuner section. As a practical note, the most likely suspect is the intercarrier sound-detector diode, which is located near the picture-detector diode on the circuit board. Associated components on the IF circuit board can also "kill" the sound signal. For example, the output coupling capacitor from the sound-detector diode may be open-circuited, or the IF bypass capacitor at the output of the sound-detector diode may be short-circuited. Signal-injection or signal-tracing tests will serve to identify the defective component, after the trouble area has been localized.

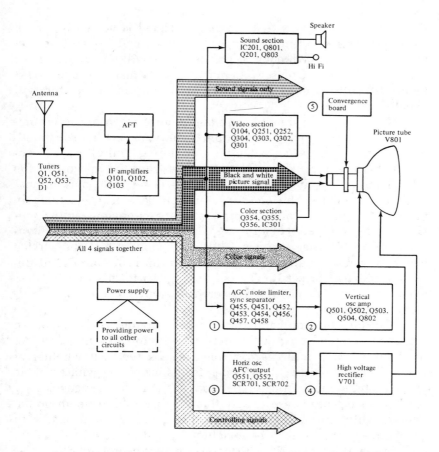

Figure 7–6. A signal flow chart. *(Redrawn by permission of Heath Company)*

Although the foregoing tests usually suffice to localize the fault when a picture-but-no-sound symptom occurs, the technician also encounters situations in which the sound signal is "killed" in the IF amplifier prior to the sound detector. This is called a separated-sound-and-picture condition. To check for this possibility, turn the fine-tuning control and observe the resulting picture and sound reproduction. Suppose that picture reception with no sound is found to occur at one setting of the fine-tuning control, and that sound reception with no picture occurs at another setting of the fine-tuning control. It is concluded that the sound signal is being trapped out completely in the IF amplifier when the receiver is tuned so that the picture signal is passed by the IF amplifier. This condition is caused by abnormal IF-amplifier action, and the IF frequency-response curve will be found to be badly distorted. Unless alignment adjustments have been tampered with, a defective bypass or decoupling capacitor will usually be found in the IF network.

Note that a separated-sound-and-picture symptom is accompanied by a separated Y-and-chroma symptom. In this separated-Y-and-chroma condition, when the fine-tuning control is set to obtain a sharp picture, the color disappears from the image. On the other hand, when the fine-tuning control is set to display normal color in the image, the picture definition becomes objectionably poor. Although this trouble symptom is not as obvious as the separation of sound and picture signals, it serves as a useful confirmation of the first analysis. Thus, a general rule at the service bench is to reserve alignment procedures until all component defects have been corrected. The only exception to this rule is taken when it is known that the viewer has tampered with the alignment adjustments.

Consider the comparatively detailed trouble chart shown in Fig. 7–7. As noted previously, if there is sound output but either no picture or an unsatisfactory picture, observe next whether or not a raster is present. If a raster is displayed, possible trouble areas include the vertical-deflection section; the pincushioning section is suspect in cases where there is distorted, little, or no deflection. On the other hand, if the raster geometry is normal with no picture visible, it may be concluded that there is a fault in the picture-signal channel. Unsatisfactory picture reproduction can be localized to the chroma section, or to both the Y and chroma sections. Picture trouble symptoms include weak or incorrect colors, misconvergence, smeary image, "snow," loss of color sync, loss of black-and-white horizontal or vertical sync, and poor focus. Note that if there is a dark screen, a picture signal will usually be present, as will be shown by a waveform check at the output of the Y amplifier.

Note that if there is a dark screen and sound output is normal, the technician knows that the trouble will not be found in the power supply. On the other hand, if there is no raster and no sound, the trouble is likely to be located in the power supply. The most probable cause of no raster and no sound is failure of one or more of the power-supply voltages. Referring to Figs. 7–8 and 7–9, the technician proceeds systematically to localize the trouble area

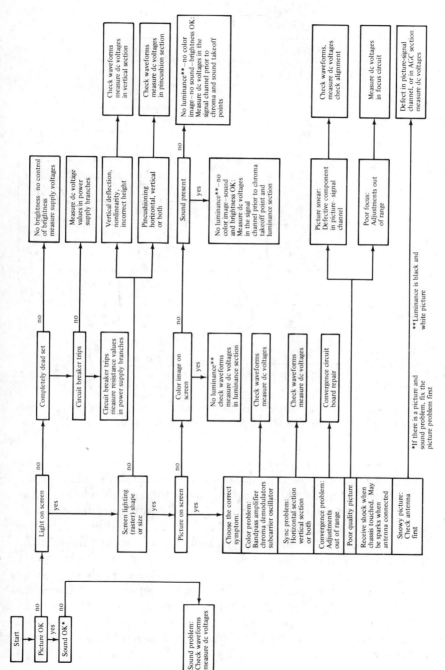

Figure 7–7. A detailed trouble-area chart. (Redrawn by permission of Heath Company)

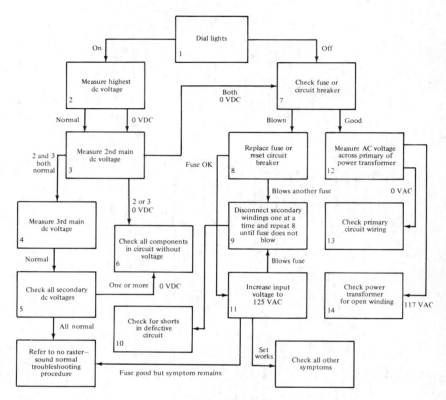

Figure 7–8. A no-raster no-sound troubleshooting chart. *(Redrawn by permission of B & K Mfg. Co. Division of Dynascan Corporation)*

as tabulated. On the other hand, if there is no raster with normal sound output, the technician follows the troubleshooting procedure listed in Fig. 7–10. This picture symptom occurs whenever the picture tube does not have normal electron-beam currents to illuminate the screen. In turn, the trouble could be caused by a defective picture tube, or by incorrect terminal voltages at the picture-tube socket. In many cases, this trouble symptom results from lack of high voltage on the anode of the picture tube. High voltage, in turn, is dependent upon the horizontal-sweep system. There are various stages in which the fault could be located; therefore, an orderly procedure is required to locate the defective component within a reasonable period of time. A recommended procedure is as follows:

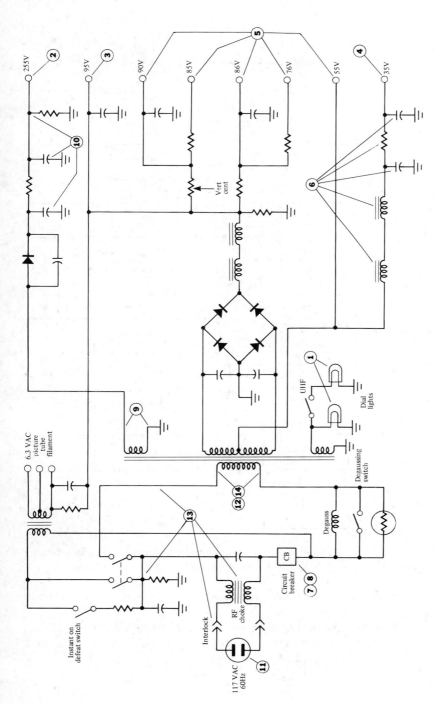

Figure 7-9. Typical power-supply schematic diagram. (Redrawn by permission of B & K Mfg. Co. Division of Dynascan Corporation)

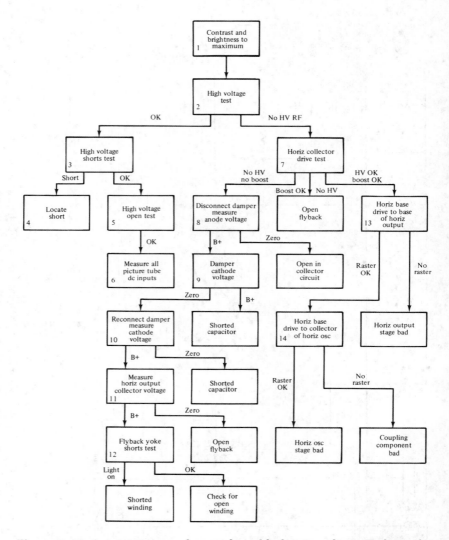

Figure 7–10. A no-raster sound-normal troubleshooting chart. *(Redrawn by permission of B & K Mfg. Co. Division of Dynascan Corporation)*

1. Set the brightness and contrast controls of the receiver to maximum.
2. Disconnect the high-voltage lead from the HV rectifier. Measure the high-voltage value. If this voltage is normal, the sweep-and-high-voltage section is cleared from sus-

picion, and attention is turned to the picture tube. On the other hand, subnormal or zero voltage indicates that the fault is located in the flyback transformer section or the high-voltage section.

3. Reconnect the high-voltage lead to the HV rectifier. If the high-voltage value decreases or goes to zero, there is evidently a short circuit present, and step 4 applies. However, if the high-voltage value is normal, proceed to step 5.

4. Next, disconnect the high-voltage lead from the picture-tube anode. If the high voltage is restored, there is an internal high-voltage short circuit in the picture tube. On the other hand, if the high voltage remains subnormal or zero, disconnect any filter capacitors, bleeder resistor, or other components, one at a time, until the high voltage reappears. The defective component is thereby pinpointed by elimination.

5. With the receiver turned off, check for a poor connection from the high-voltage lead to the anode of the picture tube. Check continuity through the high-voltage circuit. If no defect is found, proceed to step 6.

6. Check the DC voltage at the cathode and all grids of the picture tube. Investigate the reason for any abnormal or subnormal voltage.

7. Replace the connection to the high-voltage rectifier. Inject a horizontal-output drive signal at the collector of the horizontal-output transistor. If the raster display is restored, it is known that the flyback transformer and associated circuitry are workable. Thus, the defect will be found in the horizontal-sweep circuitry. In this situation, proceed to step 13 to isolate the fault. If the boost voltage appears, but the high voltage is absent, there is an open circuit in the flyback winding between the collector of the output transistor and the high-voltage rectifier terminal. If neither the boost voltage nor the high voltage is restored, there is a fault in the flyback transformer, horizontal yoke winding, or in the damper stage. In this situation, proceed to step 8. The flyback transformer and the damper stage generate the boost voltage. In the absence of boost voltage, the yoke drive will be insufficient to produce a raster.

8. Next, disconnect the damper and measure the DC anode

voltage present. This should be the supply-voltage value. If a subnormal or zero reading is obtained, measure the circuit resistances. If these are normal, proceed to step 9.

9. With the damper still disconnected, measure the DC cathode voltage. There should be no voltage at this point. If the supply voltage is measured, check for a short-circuited capacitor between the supply voltage and the boost-voltage lines. On the other hand, if no voltage is measured, proceed to step 10.

10. Reconnect the damper and again measure the cathode voltage. The supply voltage or the boost voltage should be measured at this point. If the voltage is subnormal or zero, check for a short-circuited capacitor in the cathode circuit. On the other hand, if the correct voltage value is measured, proceed to step 11.

11. Measure the DC voltage at the collector terminal of the horizontal-output transistor. The supply voltage or the boost voltage should be measured at this point. If it is zero, check for an open circuit in the flyback transformer or other series component between the damper and the horizontal-output stage. On the other hand, a normal voltage reading directs the technician to step 12.

12. Check or replace the flyback transformer and deflection yoke.

13. Replace the lead to the horizontal-output transistor and inject a base-drive signal at the horizontal-output transistor. If the raster does not appear, check the associated horizontal-output components. However, if the raster is restored, proceed with step 14.

14. A horizontal-drive signal is injected at suitable level at the collector of the horizontal-oscillator transistor. If a raster does not appear, the coupling capacitor should be checked. On the other hand, if a raster is displayed, it is indicated that the fault is in the horizontal-oscillator stage. Then make voltage and resistance checks in the horizontal-oscillator circuitry to pinpoint the defective component.

REVIEW QUESTIONS

1. How is a television analyzer used to localize faults in the signal channel?
2. Name the four signals that are considered in a flow chart.
3. What is the most likely cause for a picture-but-no-sound trouble symptom?
4. Distinguish between a no-picture and a no-raster trouble symptom.
5. Is it always possible to make a signal-tracing test in the picture channel?
6. Can a signal-injection test always be made in the picture channel?
7. Name the receiver section that is most likely to develop malfunctions.
8. If the screen is dark, how can the technician determine whether a picture signal is present?
9. Can incorrect voltages at a color picture-tube socket simulate a defective picture tube?
10. Why is a color picture-tube test jig useful in preliminary analysis of display malfunctions?
11. Is it possible for high-voltage output to be present in the absence of horizontal-scanning action?
12. Can high voltage be supplied to a picture tube from a television analyzer?
13. What is the advantage of a systematic troubleshooting approach to an obscure malfunction?
14. How is a flyback transformer checked?
15. Explain how a deflection yoke can be checked.

Troubleshooting Color Picture Tube Circuitry

8.1 GENERAL CONSIDERATIONS

Both troubleshooting and maintenance procedures are involved in correcting malfunctions of the color picture tube circuitry. This circuitry can be divided into signal, bias, and convergence sections. The signal section may include only the cathodes of the color picture tube, as shown in Fig. 8–1. On the other hand, the signal section may include both the cathodes and the grids of the picture tube, as depicted in Fig. 8–2. Preliminary tests in the signal section are made with the oscilloscope to check the drive waveforms at the signal electrodes of the picture tube. As an illustration, if a flying-spot scanner were being used to display the letter H on the picture-tube screen, the technician would normally observe the signal waveforms illustrated in Fig. 8–1. Again, if a keyed-rainbow generator were being used to display a chroma-bar pattern on the screen, pulse waveforms would normally be observed, as explained in the preceding chapter.

Two principal types of color picture tubes are in general use. These are described as the shadow-mask design and the aperture-grille design. Most color receivers in present use have shadow-mask picture tubes. There are various sizes and minor design differences in shadow-mask tubes, but all have comparatively elaborate convergence circuitry. There are also various sizes and design differences in aperture-grille tubes. However, most aperture-grille tubes

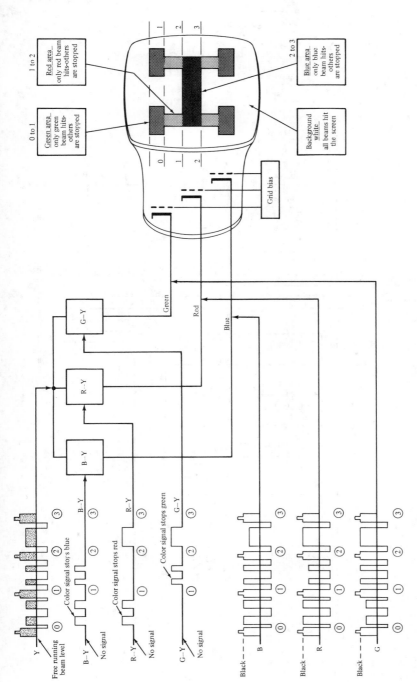

Figure 8–1. An example of cathode drive to the cathode ray tube. (Redrawn by permission of Heath Company)

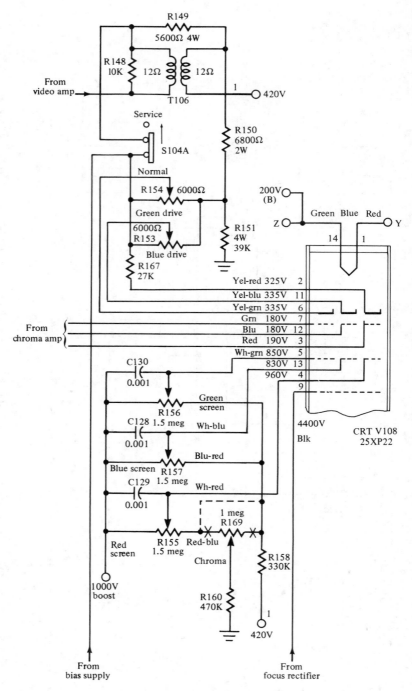

Figure 8–2. An example of cathode and grid drive to the color picture tube.

have comparatively simple convergence circuitry. One design of aperture-grille tube has unadjustable (fixed) convergence facilities, so that the technician is never concerned with converging the picture tube. This type of color picture tube has its deflection and convergence components permanently bonded to the neck of the tube. Accordingly, when the picture tube is replaced, the deflection and convergence components are discarded with the defective picture tube.

Figure 8–3 shows the terminal points and components on a conventional color picture tube. There are two heater terminals, three cathode terminals, three control-grid terminals, three screen-grid terminals, a focus-grid terminal, and a high-voltage anode terminal brought out from the picture tube. External component circuit terminals include four deflection-yoke connecting leads and 12 convergence-coil leads. DC voltage is applied to all of the picture-tube terminals, with the exception of the heater, which op-

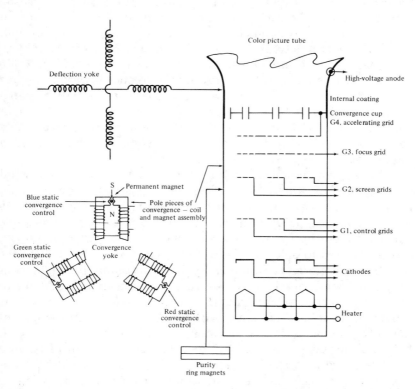

Figure 8–3. Terminal points and components on conventional color picture tube.

erates on AC voltage. The deflection-yoke windings and the convergence-yoke windings are driven by nonsinusoidal AC voltages. Convergence circuitry is comparatively complex, and employs inductors, capacitors, resistors, and semiconductor diodes. Capacitors are generally checked first in case of malfunction, and diodes are the next most likely components to develop defects.

8 . 2   SERVICING THE CONVERGENCE CIRCUITRY

Dynamic convergence adjustments are made to obtain a properly converged white-dot or crosshatch pattern over the entire screen area of the picture tube. Dots at the center of the screen are brought into convergence by adjustment of the static convergence controls. As the dynamic-convergence adjustments are made, convergence at the center-screen area may tend to become "pulled out." Therefore, whenever misconvergence becomes evident in the central screen area, the static convergence controls are touched up, as required. Figure 8–4 depicts a typical dynamic convergence-control board layout. These controls are adjusted in sequence, as tabulated in Table 8–1, to obtain progressively improved convergence toward the edges of the screen. In most cases, the screen convergence will be satisfactory after completion of the fourteenth adjustment. However, if the initial convergence is very poor, it may be necessary to repeat part or all of the dynamic-convergence adjustments, owing to control interactions.

Some shadow-mask picture tubes require blue horizontal shaping-coil adjustment, wide blue-field adjustment, and dynamic pincushion adjustments. When a blue horizontal shaping coil is provided, a test point is available on the dynamic-convergence panel for connection of an oscilloscope. The slug in the shaping coil is adjusted to obtain the waveform depicted in Fig. 8–5(f). Unless this waveform is properly shaped, optimum dynamic convergence cannot be obtained, because the blue lines will tend to be curved instead of straight. If a wide blue-field adjustment is provided, an adjusting screw will be observed on the bottom of the yoke assembly. This adjustment is made to bring the vertical height of the blue field exactly the same as the height of the red and green fields. Misadjustment of the blue-field height will impair dynamic convergence. Pincushion distortion is depicted in Fig. 8–6. Dynamic pincushion adjustments are commonly pro-

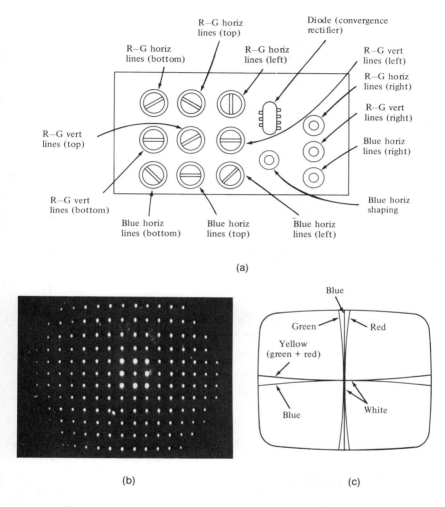

Figure 8–4. Dynamic convergence controls for conventional picture tube:
(a) Control-board layout; **(b)** Dot pattern displayed for good convergence;
(c) Good static convergence; poor dynamic convergence.

vided for large-screen shadow-mask picture tubes. These adjustments are made at the factory and seldom require attention. However, if the raster edges appear curved instead of straight, the
pincushion adjustments should be touched up as required. This
topic is detailed subsequently.

TABLE 8–1

CONVERGENCE ADJUSTMENTS

Step	Control	Use to converge (or straighten)	Remarks
1.			Perform center dot convergence using convergence magnets. (See Fig. 8-5 (a))
2.	R–G vertical lines, top	Red and green vertical bars at top of screen	Touch up both controls for best convergence from top to bottom along vertical center line (Fig. 8-5 (b))
3.	R–G vertical lines, bottom	Red and green vertical bars at bottom of screen	
4.	R–G horizontal lines, top	Red and green horizontal bars at top of screen	Touch up both controls for best convergence of horizontal bars along vertical center line (Fig. 8-5 (b))
5.	R–G horizontal lines, bottom	Red and green horizontal bars at bottom of screen	
6.	Blue horizontal lines, top	Blue horizontal bars at top of screen	Touch up both controls for best convergence of horizontal bars along vertical center line (Fig. 8-5 (c))
7.	Blue horizontal lines, bottom	Blue horizontal bars at bottom of screen	
8.			Perform center dot static convergence (Fig. 8-5 (a))
9.	Blue horizontal lines, right	Blue horizontal bars at right side of screen	Touch up both controls for best convergence along horizontal center line (Fig. 8-5 (d))
10.	Blue horizontal lines, left	Blue horizontal bars at left side of screen	
11.	R–G vertical lines, right	Red and green vertical bars at right side of screen	(Fig. 8-5 (e))
12.	R–G horizontal lines, right	Red and green horizontal bars at right side of screen	Use control to converge blue bar with red and green bars on right side of screen (Fig. 8-5 (e))
13.	R–G vertical lines, left	Red and green vertical bars at left side of screen	
14.	R–G horizontal lines, left	Red and green horizontal bars at left side of screen	Use control to converge blue bar with red and green bars at left side of screen (Fig. 8-5 (e))

8 . 3 DYNAMIC CONVERGENCE CIRCUIT ACTION

Troubleshooting the dynamic-convergence circuitry requires an understanding of the circuit actions that are associated with trouble symptoms. As noted previously, edge-screen convergence is effected by horizontal and vertical electromagnets mounted over the electron guns on the picture-tube neck. Current waveforms in step with the horizontal- and vertical-sweep voltages are passed through the horizontal and vertical dynamic-convergence coils (Fig. 8–7). These electromagnetic fields correct the beam directions before they are deflected by the yoke fields. Dynamic-convergence waveforms necessarily have a critical amplitude, shape, and phase. Basically, the corrective current waveform that is required has the shape of a parabola, as depicted in Fig. 8–7(b).

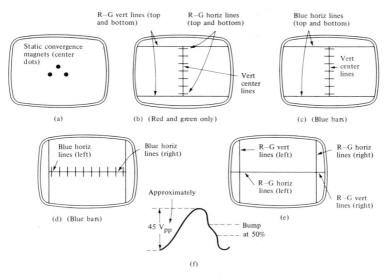

Figure 8–5. Convergence patterns and terminology: **(a)–(e)** Static and dynamic convergence indications; **(f)** Shaping-coil waveform.

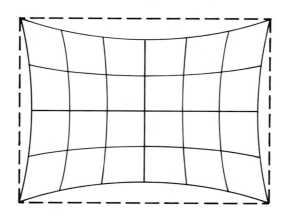

Figure 8–6. Curved vertical and horizontal lines caused by pincushion distortion.

Vertical parabolic current waveforms are obtained by processing the vertical sawtooth deflection voltage. Partial integration of this sawtooth waveform produces an approximation of a para-

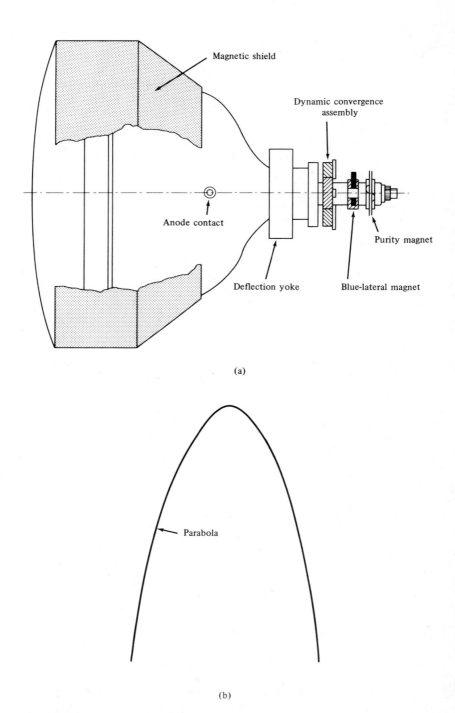

(a)

(b)

Figure 8–7. Dynamic convergence assembly: **(a)** Location on picture-tube neck; **(b)** Basic current waveform is a parabola.

bolic waveform. A sectional schematic diagram is shown in Fig. 8–8. Current flowing through C52, R51, differential resistor R52, the red and green convergence coils L4 and L5, diode D53, and to the B+ supply converges the upper half of the raster. Note that capacitor C52 and resistors R54 and R55 operate in a waveshaping circuit that produces the required convergence waveform. Current flowing through resistor R54, differential resistor R53, the convergence coils, amplitude resistor R58, shaping-network diodes D51 and D52 and their associated capacitor, and resistor R56 with C51 converges the lower half of the raster. This second circuit operates during the time that D53 is not conducting.

Observe that the amplitude controls in Fig. 8–8, R54 and R58, control the amount of correction in convergence of the vertical lines, whereas the differential controls R52 and R53 govern the amount of correction in vertical convergence of the horizontal lines at the top and bottom of the raster. Figure 8–9 shows the basic convergence circuit for vertical convergence of the blue horizontal lines. Controls R59 and R60 affect convergence at the top and bottom halves of the raster. This circuit is energized from the same source that is utilized by the red and green convergence circuits. However, its output is supplied to L6, the convergence coil mounted over the blue gun.

Next, consider the plan of the horizontal convergence circuitry. Figure 8–10 shows a basic schematic diagram of this section. Flyback pulses are fed to coil L51, controls R67 and R68, and capacitors C59 and C60 to generate sawtooth waveforms that are applied to the red and green convergence coils. In turn, integrating action produces parabolic current waveforms through the coils. Note that clamping diodes D55 and D56 and the associated resistors rectify a portion of the parabolic current waveform, thereby adding a DC component to the convergence waveform. This DC component is required to ensure that the current through the convergence coils is zero as the scanning beam passes through the center of the screen. Observe that coil L54 and control R67 provide a difference-current flow through the red and green convergence coils. This differential current flow corrects symmetrical errors in horizontal-line convergence on the left and right sides of the screen.

Figure 8–11 shows a basic configuration for the blue horizontal convergence circuits, which operate in basically the same manner as the green and red horizontal convergence circuits previously described. However, an additional wave-shaping network

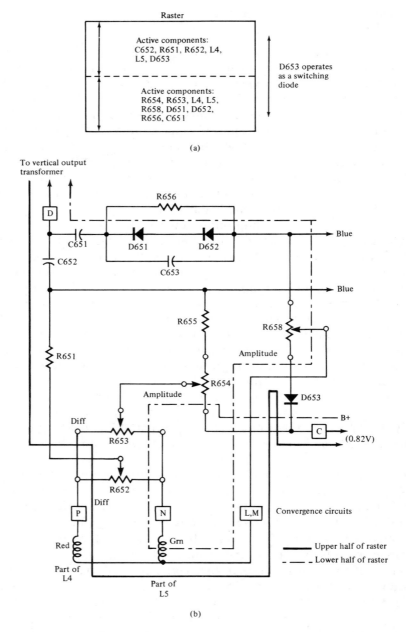

Figure 8–8. Vertical parabolic waveforms are produced for convergence of the upper and lower portions of the raster by integrating a sawtooth waveform: **(a)** Portions of raster that are separately converged; **(b)** Network arrangement.

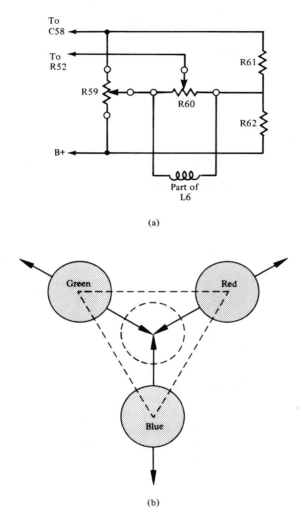

(a)

(b)

Figure 8–9. Basic convergence circuit for vertical convergence of the blue horizontal lines: **(a)** Configuration; **(b)** Dot motions produced by convergence fields.

comprising coil L53, capacitor C57, and resistor R63 is included. This waveshape adds a small amount of second-harmonic sine-wave current to the blue-convergence coil current. Thereby, the blue horizontal convergence waveform is optimized.

A complete convergence circuit-board configuration is shown

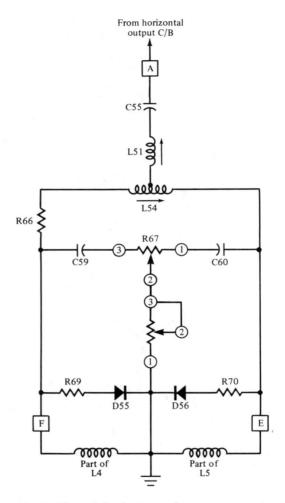

Figure 8–10. Plan of the horizontal convergence circuitry.

in Fig. 8–12. This assembly employs the partial circuit diagrams, with the same identifications, that have been discussed. Note that 13 maintenance controls are provided. Nine of these are resistive controls; four are inductive controls. As anticipated, there is considerable interaction among the convergence controls, because the fields from the three pole pieces affect the directions of all three electron beams to a greater or lesser extent. In practice, this interaction results in a comparatively involved convergence setup

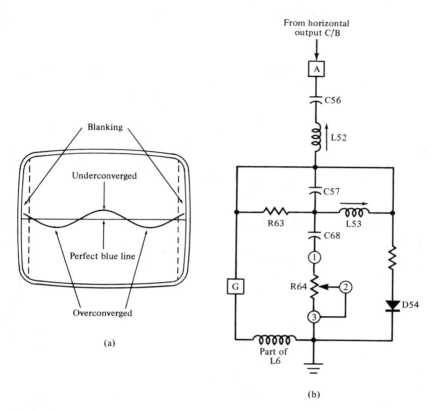

Figure 8–11. Effect of second-harmonic voltage in blue horizontal convergence: **(a)** Underconverged, overconverged, and perfect blue-line displays; **(b)** Basic circuitry for blue horizontal-convergence section.

procedure. Skill and speed in the convergence of a shadow-mask picture tube are acquired by a combination of study and experience.

In the example of Fig. 8–12, six other maintenance controls are mounted on the convergence panel assembly. However, these are not included in the convergence circuitry. Thus, the dot control operates in conjunction with the horizontal-sweep circuit to generate white-dot pulses that are utilized in convergence checks. (Note that most color receivers do not have built-in dot generators.) The height control is part of the vertical-deflection system; the AGC control is part of the gain-control system; the tone control is part of the sound system; the color-killer control is part of

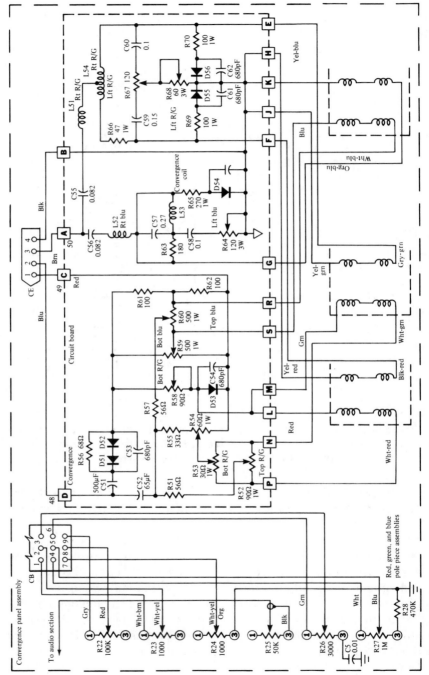

Figure 8-12. Configuration for a convergence circuit board.

the chroma system. The video peaking control provides an adjustable load resistance so that the high-frequency response in the picture channel can be trimmed to optimize weak-signal or strong-signal reception.

8.4 APERTURE-GRILLE PICTURE TUBE

Convergence procedures are considerably simplified in the case of most aperture-grille picture tubes, and are somewhat simplified even with the elaborate designs. A few designs practically eliminate convergence adjustments. With reference to Fig. 8–13, the basic components of the aperture-grille color picture tube are the electron-gun assembly, the aperture grille, and the phosphor-stripe screen. Note that the three electron guns are located in line horizontally. All three electron beams pass through the same slot in the grille, but at different angles. In turn, the beam from the red gun strikes the red phosphor stripe, the beam from the green gun strikes the green phosphor stripe, and the beam from the blue gun strikes the blue phosphor stripe. This action is normally the same for any slot at any point on the screen.

An aperture-grille picture tube operates from the same signal waveforms as the shadow-mask tube. As an illustration, if the red and green guns were energized, and the blue gun cut off, a yellow field would be displayed with either type of picture tube. Again, if all three guns were energized, a white field would be displayed with either type of picture tube. On the other hand, the convergence waveforms and adjustments for the two types of picture tubes are quite different. As shown in Fig. 8–13, the aperture-grille picture tube employs three in-line cathodes. Emitted electrons pass through small holes in the control grid G1. Next, the electrons are speeded up by the screen grid or accelerating electrode, G2. In turn, the electron beams pass through holes in the electrode and proceed at high speed into the focus-electrode region. This is essentially an electrostatic-lens assembly, which brings the electron beams to a point on the aperture grille. Only the green electron beam proceeds in a straight line. The red and blue beams come together at a point and then diverge. This focusing action also has the effect of minimizing the diameter of each electron beam. These three beams are attracted to the converging plates, which operate typically at 19,000 volts. Thus, a second electron-

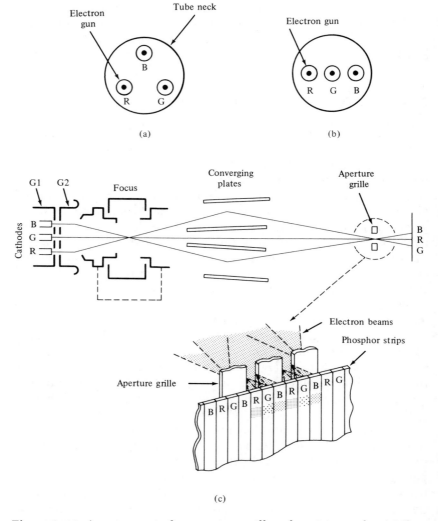

Figure 8–13. Arrangement of an aperture-grille color picture tube: **(a)** Conventional delta gun arrangement; **(b)** In-line gun arrangement; **(c)** Functional features of the in-line tube.

lens arrangement brings the beams to precise focus on the aperture grille.

Field purity is obtained by proper adjustment of the purity magnet and by correct positioning of the deflection yoke. With reference to Fig. 8–14, the purity magnet is adjusted to obtain opti-

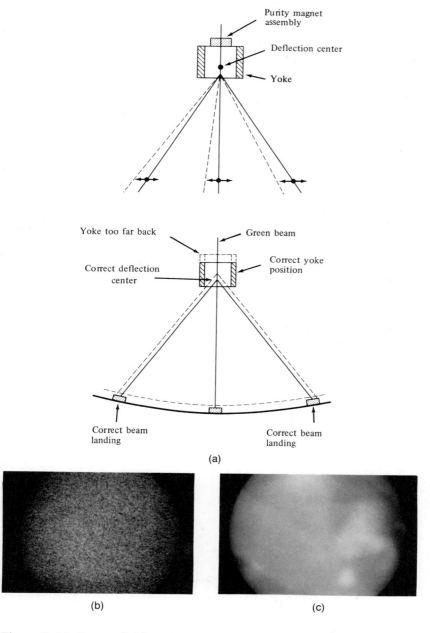

(a)

(b)   (c)

Figure 8–14. Correct field purity requires proper positioning of the deflection yoke: **(a)** Principle of purity adjustment; **(b)** Good field purity; **(c)** Poor field purity.

mum purity at the center of the screen. Then the deflection yoke is moved to a position that provides maximum purity out to the screen edges. A green field is ordinarily used in this procedure. Note that the purity magnet shifts all three beams by the same amount at all points on the screen. On the other hand, the deflection-yoke position shifts the beams much more at the screen edges than at the center of the screen. Another purity control utilized on the aperture-grille tube is called a *neck-twist coil*. It is mounted at the rear of the tube neck. This adjustment affects the red and green beams only, as depicted in Fig. 8–15. Basically, this arrangement is the same purity magnet used with a shadow-mask type of tube. Adjustment of the neck-twist coil is usually made with a red field display.

Convergence at the edges of the screen requires the addition of a parabolic voltage waveform to the converging plates in Fig. 8–13. In other words, a parabolic waveform (Fig. 8–16) is superimposed on the DC focusing voltage. This waveform can be tilted one way or the other by means of a sawtooth voltage from the horizontal dynamic control. This tilt is provided to compensate for manufacturing tolerances on replacement picture tubes. The final convergence adjustment control is called a vertical static control. This is a DC electromagnet arrangement that operates as a static convergence control (see Fig. 8–17). This coil is mounted so that the vertical beam directions are affected. It is the only vertical convergence control provided for the aperture-grille type of picture tube, and it is generally adjusted first in the convergence procedure.

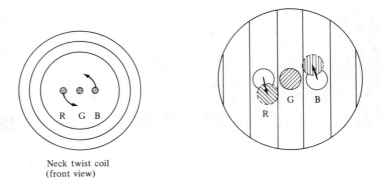

Neck twist coil
(front view)

Figure 8–15. Principle of neck-twist coil action.

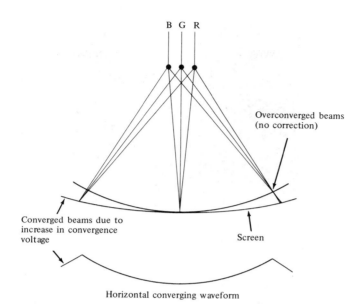

Figure 8–16. Electron beams are over-converged at the screen edges unless a parabolic correction voltage is utilized.

Observe in Fig. 8–17 that the cathodes are driven by the outputs from the color video amplifiers. In other words, the aperture-grille tube is not used as an RGB matrix. All matrixing operations are accomplished in the receiver circuitry. Neon lamps are provided in the cathode circuits to protect the picture tube against possible excessive cathode potentials. R1 and C1 provide a compensated attenuating pad, which operates as part of the video-amplifier system to provide correct relative color-signal levels. As noted previously, the readjusted chroma values that energize the chroma demodulators must be changed into unadjusted chroma values before application to the color picture tube. Otherwise, the relative color intensities would be in error, and the blue hues in particular would be too weak. Blanking pulses are applied to the control grid of the picture tube to make the retrace lines invisible.

From a practical viewpoint, the in-line tube is much the same as the trinitron picture tube, except that no convergence adjustments are required. An in-line picture tube has factory-sealed purity and convergence components, and requires no adjustment. The deflection yoke is also permanently mounted on the neck of

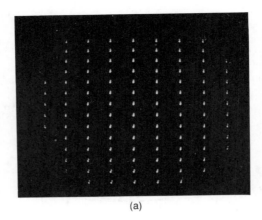

(a)

Figure 8–17. The vertical static control is an electromagnetic field adjustment: **(a)** Example of misadjustment;

the tube. In turn, when the picture tube is replaced, the yoke, purity, and convergence components must also be replaced. Figure 8–18 shows the configuration of the horizontal-sweep and power-supply circuits used with the in-line picture tube. Note that the trace and retrace diodes are inside the same cases as the trace and retrace SCR's. These components are called intrinsic rectifiers (ITR's). No regulator transistor, saturable reactor, or high-voltage adjustment is used in this horizontal-sweep and high-voltage section. A solid-state tripler is employed to rectify the high-voltage waveform.

Most of the power-supply voltages are obtained from the horizontal-sweep system depicted in Fig. 8–18. Only the horizontal-deflection circuitry obtains power by rectification directly from the 117-volt AC line. Approximately 150 volts is supplied by the half-wave rectifier at all times, and the horizontal oscillator and retrace ITR operate continuously, regardless of whether the receiver is turned on or off, or is in the standby condition (effectively off). Output from the horizontal oscillator is a series of pulses at a 15,750-Hz rate, each pulse having a width of approximately 6 microseconds. Each pulse momentarily keys on the retrace ITR. In the standby condition, therefore, the small ITR current through T402 provides heater current for the picture tube. This is called "instant-on" operation.

There is no drive to the yoke or the flyback transformer during the standby condition, because both of these components are short-circuited by a section of the on-off switch (anode of ITR1 is switched

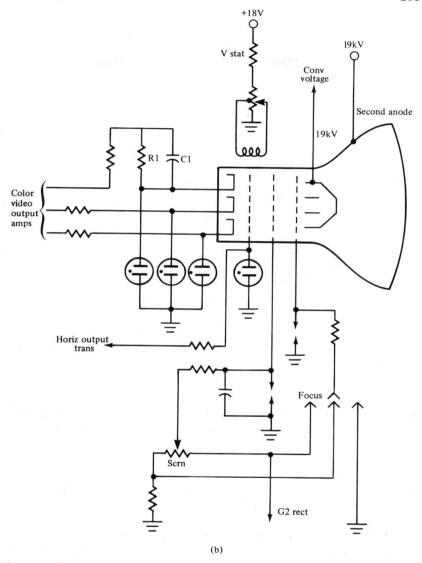

(b)

Figure 8–17 continued. (b) Configuration

to ground). Note that there are two other sections in the power switch. One section disconnects the 40-volt supply from its rectifier, CR5. The other section connects the degaussing coil to the AC line. Provision of a built-in degaussing coil ensures that the picture tube

(a)

Figure 8-18. Horizontal sweep and power supply system for an in-line picture tube: (a) Oscillator output waveform; (b) System configuration.

or its supporting structures will never become magnetized and impair the screen purity. Degaussing takes place immediately following turn-on. Soon the thermistor RT3 reduces the degaussing current to a negligible value.

When the receiver is turned on, the short circuit is removed from across the yoke and flyback transformer, and the degaussing coil is disconnected. At this time, deflection current and high-voltage output are generated. This action increases the load-current pulses through T2, thereby increasing the heater voltage for the picture tube to normal value. At the same time, CR5 is switched into its loads. The 40-volt supply is filtered and fed to the vertical-sweep system, and voltage dividers energize the 30-volt buses to the signal modules. (Module servicing is explained in greater detail in the next chapter.) Note that a screen-grid voltage of approximately 850 volts, and 220 volts for the video-drive modules are rectified from taps provided on a flyback transformer winding. The focus voltage is taken from a tap on the high-voltage tripler assembly.

Troubleshooting of this circuitry is not unduly difficult, although the specialized design of the power supply involves an important precaution. If serious trouble occurs in the horizontal-sweep system, the remainder of the receiver will be "dead," because of the associated power-supply failure. In other words, the technician should check the supply voltages before assuming that there are defective components in any receiver section. A receiver that utilizes an in-line picture tube should not be energized from a wall outlet that is switched. Not only would this defeat the "instant-on" feature of the receiver, but the picture tube could possibly be damaged from application of high voltage before the cathodes warm up and start emitting electrons.

REVIEW QUESTIONS

1. What are the subdivisions encountered in color picture-tube circuitry?
2. Name the two principal types of color picture tubes.
3. Explain the function of the dynamic convergence section.
4. How does a wide blue-field adjustment function?
5. Define pincushion distortion.
6. Describe the development of a parabolic dynamic-convergence waveform.
7. State the number of convergence controls provided in a conventional dynamic-convergence system.
8. Why do many dynamic-convergence adjustments tend to interact?
9. What is the basic distinction between a shadow-mask picture tube and an aperture-grille picture tube?
10. Does an aperture-grille picture tube require a different video-signal input, compared with a shadow-mask picture tube?

11. Can an aperture-grille picture tube be utilized as an RGB matrix?

12. How does an ITR differ from an SCR?

13. What is the function of a degaussing coil?

14. Explain the function of a thermistor in a built-in degaussing arrangement.

15. Why should an instant-on type of receiver never be operated from a wall outlet that is switched?

Modular Troubleshooting Procedures

9.1 GENERAL CONSIDERATIONS

Various designs of color-TV chassis have modular construction. Modules are plug-in circuit-board units, such as those illustrated in Fig. 9-1. This design tends to simplify troubleshooting procedures, because a module with a defective component can be replaced with a new (or reconditioned) module, and the receiver can thereby be restored to normal operating condition promptly. In turn, the defective module can be repaired at any time. If a module is seriously damaged, the technician may choose to discard it. A color receiver that has modular construction is not taken to the shop for repairs—all servicing procedures are done in the home.

A typical receiver employs nine modules to perform the functions of IF amplification and automatic frequency control, sound demodulation, video amplification and picture-tube driving, chroma demodulation, vertical deflection, and horizontal deflection. A power-supply module may also be provided in some receivers. Note that the functional classification of a module may be complete or partial. As an illustration, a module may have a complete video-amplifier function and partial sync function, such as a sync-separator stage. Again, a module may have a partial

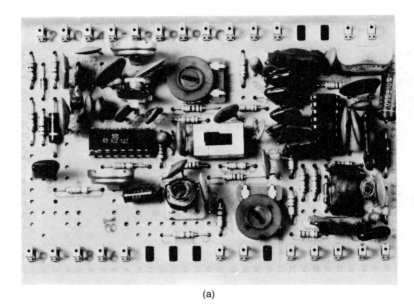

(a)

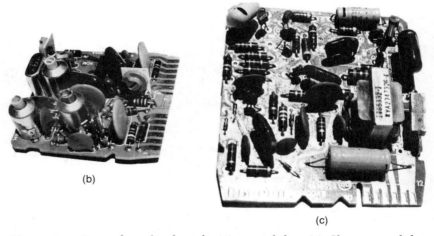

(b)

(c)

Figure 9–1. Examples of color-television modules: **(a)** Chroma module; *(Courtesy of RCA)* **(b)** Another type of chroma module; *(Courtesy of Zenith)* **(c)** A horizontal-oscillator module. *(Courtesy of RCA)*

function completed by conventionally mounted chassis components. For example, an audio-amplifier module may have to be supplemented by chassis-mounted audio-output components.

A block diagram for a modular receiver, with the functions of the various modules, is shown in Fig. 9–2. The tuner, as would be expected, is a chassis-mounted component. It is coupled by a

(a)

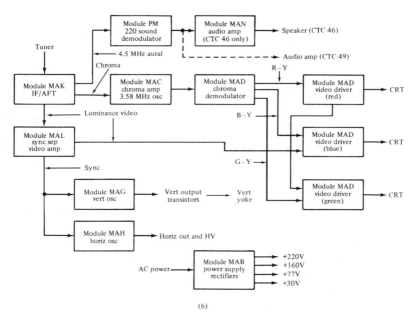

(b)

Figure 9–2. Modular chassis arrangement: **(a)** Rear view of RCA CTC-49 chassis; **(b)** Block diagram of a modular chassis.

50-ohm line to the IF module, as depicted in Fig. 9–3. This coupling method simplifies servicing procedures, because it makes the alignment of L1 independent of the alignment of L2, L3, L4, T1, and T2. In other words, if the tuner is replaced, realignment of the IF-module tuned circuits is not required. Or, if the IF module is replaced, realignment of the tuner output coil L1 is not required. Next, a simplified block diagram of an IF module is shown in Fig. 9–4. Two integrated circuits and two transistors are utilized in this example. When IF trouble symptoms occur, it is sometimes possible to pinpoint the defective component promptly and to repair the module without delay. As an illustration, if either of the transistors happens to fail, or if there is some visible defect, an easy and rapid repair may be possible. On the other hand, when the trouble cannot be localized within a few minutes, it is good practice simply to replace the module.

If picture symptoms indicate trouble in the "receiving" section, there may be a defect in either the tuner or in the IF amplifier. Accordingly, it is advisable to substitute a new IF module at the outset, to localize the malfunction. This test requires only half a minute, after the rear cover of the cabinet has been removed. However, in the event that a substitute IF module is not available, another quick check can be made as follows. With the IF module removed, a small capacitor is temporarily connected from the tuner output to the 4.5-MHz sound-input terminal. Since the 4.5-MHz intercarrier sound signal appears at the output of the VHF mixer stage, this test serves to indicate whether the tuner is operative. This is a practical test procedure because the IF amplifier normally contributes comparatively little gain for the 4.5-MHz signal, and because the intercarrier-IF amplifier has substantial reserve gain. Figure 9–5 shows how this "bridging" test is made.

A voltage-divider type of volume control is shown in Fig. 9–5. Some modules utilize an integrated circuit instead of individual semiconductor components, and have an unconventional type of volume-control action. In this design, the manual volume control may merely vary the bias of an amplifier stage within the integrated circuit and thereby vary the audio gain. In such a case, it may be falsely concluded that the integrated circuit is "dead," merely because no audio signal is present at the volume control. Conversely, an audio signal cannot be injected at this type of volume control for a signal-substitution test. In other words, the technician should not assume that modular construction employs conventional circuitry; it

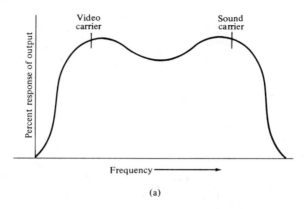

(a)

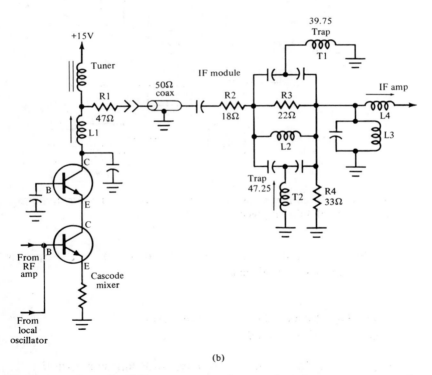

(b)

Figure 9–3. Coaxial cable arrangement for coupling the mixer stage to the IF section: **(a)** Normal RF frequency response curve; **(b)** Intersection coupling circuitry.

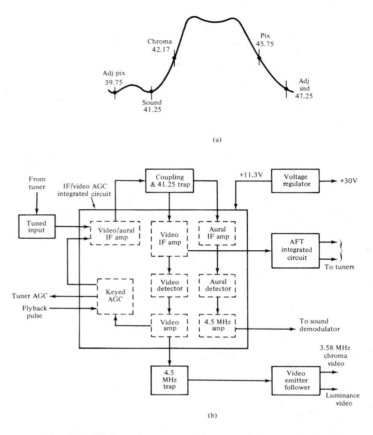

Figure 9–4. Simplified block diagram of an IF module: **(a)** Normal frequency response curve; **(b)** Subsections of module.

is sometimes essential to check the circuit diagram for the module before making operational tests.

9 . 2 TROUBLESHOOTING HORIZONTAL-DEFLECTION UNITS

Most modular color-TV receivers have SCR horizontal-deflection circuits. Troubleshooting these silicon controlled rectifier circuits is somewhat different from the procedures used in transistor deflection circuits, because of the difference in device character-

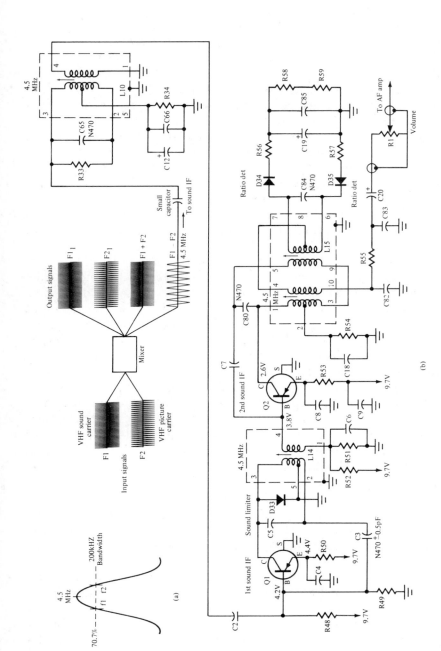

Figure 9-5. Bridging test from VHF tuner to sound-IF section: **(a)** Normal frequency response of the sound-IF section; **(b)** A 4.5-MHz signal is coupled into the sound section.

istics. An SCR belongs to the same family of solid-state devices as the triac, a gate-controlled full-wave silicon switch. These devices do not normally conduct unless they are turned on by a keying pulse, or a trigger waveform. After being turned on, they remain in their conductive state until the cathode-anode voltage is reduced to practically zero. Note that a conducting SCR cannot be turned off by applying negative bias to its gate electrode. Figure 9–6 shows the essentials of an SCR deflection system. If SCR2 or D2 "shorts," horizontal deflection stops and there is no high-voltage output. In case excessive current is drawn, and the circuit breaker trips, the technician turns his attention at the outset to SCR1 and D1, and tests them for short circuits. These same trouble symptoms can be caused by serious leakage or by a short circuit in D2 or C3. Defective components can usually be pinpointed by means of DC-voltage measurements, compared with specified values in the receiver service data.

When an open circuit occurs anywhere in the load-circuit branches of the configuration in Fig. 9–6, a more demanding troubleshooting situation is encountered. The most serious open-circuit condition involves the horizontal-deflection yoke winding. This fault usually results in excessive voltage rise across SCR2 and D2 with the result that one or both of the devices are likely to be burned out. Whenever SCR2 or D2 is replaced because of an open- or short-circuit fault, it is good practice to determine the reason for the failure, and to correct it. Otherwise, there is good probability that the replacement device will soon burn out. First, the technician checks the continuity of the yoke and flyback primary windings. As would be expected, intermittent "opens" are most troublesome when the symptom is repeated failure of D2 and SCR2. As an illustration, D2 may be loose in its retaining clips; there may be loose connections in the yoke socket; screws that hold SCR2 in its socket might be loose; there could be a cold-solder joint where T1 is soldered to the circuit board; sometimes a pinhole forms through the mica insulator between SCR2 and its heat sink, permitting an occasional arc, which destroys the semiconductor junction.

Loss of horizontal sync is another type of general trouble symptom. When this malfunction is localized to the SCR deflection circuit, the most likely cause is an interchange of SCR1 and SCR2, or D1 and D2 (see Fig. 9–7). In other words, if these devices become accidentally interchanged during servicing procedures, the

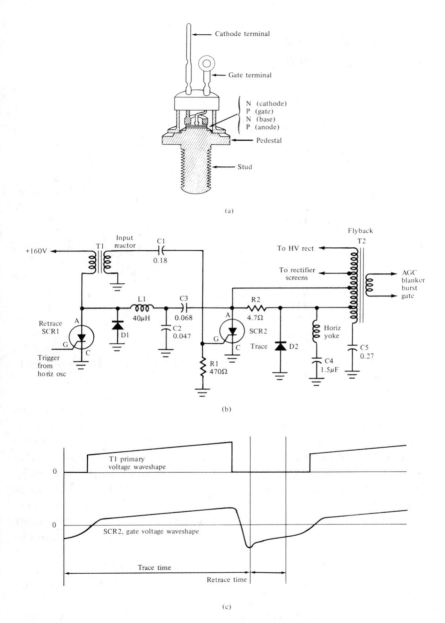

Figure 9-6. Essentials of an SCR deflection system: **(a)** Construction of an SCR; **(b)** Basic horizontal-deflection circuitry; **(c)** Gate waveforms in the SCR2 branch.

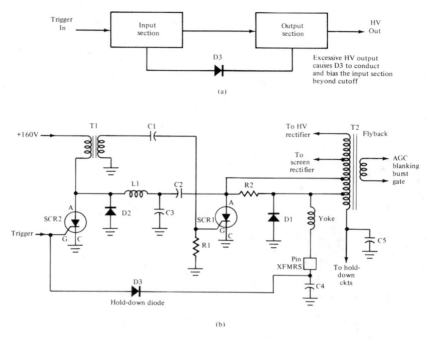

Figure 9–7. SCR deflection system, with hold-down diode: **(a)** Functional block diagram; **(b)** Configuration.

system will almost work normally, but not quite. Also, note the block marked "pincushion transformers" in the diagram. Figure 9–8 shows pincushion correction circuitry. Its function is to compensate for curvature in the right-hand and left-hand edges of the raster, and along the top and bottom, so that the raster has straight edges. If C4 becomes "open," the voltage across SCR2 and D2 rises considerably, and these devices are likely to be damaged. Similarly, other defects that interrupt the circuit from yoke to ground through the pincushion correction circuitry will cause this trouble symptom.

Observe in Fig. 9–7 that a lead from the lower end of T2 connects to a network called a hold-down circuit. Other designations are fail-safe and high-voltage runaway protective circuit. It functions to disable the horizontal-sweep system if a fault occurs that causes the high-voltage value to rise excessively. Note that the hold-down diode D3 is normally reverse-biased, because the voltage drop across C4 is always positive. On the other hand, if D1

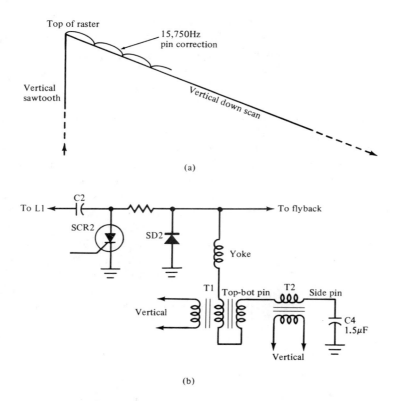

Figure 9–8. Pincushion corrective network: **(a)** Sawtooth waveform with pincushion correction; **(b)** Configuration.

should become defective (open-circuited), the voltage across C4 swings negative part of the time and forward-biases D3. The anode of D3 is returned to the gate of SCR2. In normal operation, D3 is effectively out of the circuit. However, in a trouble situation that causes D3 to conduct, SCR2 becomes biased beyond cutoff and the sweep section is disabled. This cutoff voltage decays, and the sweep section starts to operate again, only to be cut off once more. In turn, a loud squeal is heard from the transformers; the raster is shrunken and torn, and the circuit breaker may trip.

With reference to Fig. 9–8, top and bottom pincushion correction is obtained by passing the current from the horizontal yoke winding through T1, which energizes the 15,750-Hz resonant circuit that returns to ground via C4. In turn, each horizontal scan generates a half-sine wave that is added to the vertical-yoke waveshape in order to compensate for top and bottom pincushion distortion. Also, side pincushion correction is obtained by 60-Hz modulation of the horizontal current sawtooth waveform. Saturable reactor (nonlinear inductor) action is provided by T2 to develop an increase in horizontal width over the central portion of the raster, thereby compensating for side pincushion distortion.

Next, consider the hold-down circuit depicted in Fig. 9–9, which will be encountered in various modular receivers. As in the arrangement discussed previously, this hold-down circuit has the function of inhibiting steady operation of the horizontal-sweep system in the event of a component failure that causes excessive high voltage to be generated. In typical modular receivers, the anode of the hold-down diode is connected into the circuit of the horizontal-hold control, as shown in this example. Also, a second hold-down circuit is employed in this arrangement. It supplements the function of the hold-down diode for some component defects that would be inadequately controlled otherwise. This is the function of Q402. With reference to Fig. 9–7, this second hold-down circuit is connected at the return end of the flyback transformer T2.

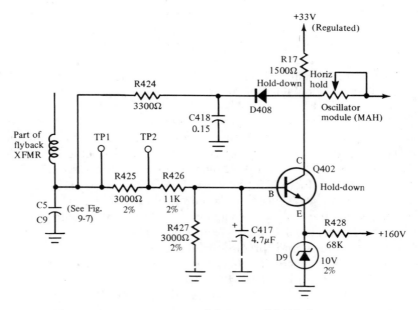

Figure 9–9. Arrangement of the second hold-down circuit.

Observe in Fig. 9–9 that development of excessive high voltage will cause the peak drop across C4 and C5 to rise. In turn, Q402 starts to conduct sufficiently that the voltage is decreased on the high end of the horizontal-hold control. This voltage change shifts the horizontal-oscillator frequency out of sync, and to an abnormally high frequency, thereby reducing the system efficiency and attenuating the high voltage. From a troubleshooting procedural standpoint, an excessive horizontal-scanning frequency could also be caused by a defect in the horizontal AFC section or in the horizontal-oscillator circuit. It could even be caused by a defect in the hold-down circuit itself. Therefore, localization tests are required. Check first to make certain that the coarse-hold control in the horizontal-oscillator circuit has not been misadjusted. With the horizontal-hold control set to the midpoint of its range, it is normally possible to synchronize the horizontal oscillator by adjustment of the coarse-hold control. However, if this adjustment does not provide sync lock, the technician concludes that there is a component defect to be tracked down.

Preliminary localization is made by grounding the test point marked TP2 in Fig. 9–9. If the horizontal oscillator can now be brought on frequency, it is logical to conclude that the trouble will be found in the high-voltage regulator section. On the other hand, if the horizontal-oscillator frequency cannot be corrected, the technician looks for a defect in the hold-down circuitry, or in the oscillator section. This possibility is easily checked by replacing the oscillator module. If the hold-down system comes under suspicion, the collector of the hold-down transistor is disconnected and the anode of the hold-down diode is disconnected. These tests usually localize the trouble to a specific section, after which DC-voltage and resistance measurements are made to pinpoint the defective component. Or the module with the defect may be simply replaced, and subsequently repaired or discarded.

Inasmuch as an SCR deflection system involves some specialized troubleshooting procedures, it is helpful to summarize the following basic facts:

1. Although the system becomes "dead" if the horizontal-oscillator drive waveform is lost, the components in an SCR deflection system will not be damaged.
2. Component damage can be anticipated if the yoke circuit becomes "open" for any reason. Therefore, whenever an

SCR or its associated diode is replaced, the technician should check the continuity of the yoke circuit before attempting to operate the receiver.

3. Although a sustained overload is likely to damage the flyback transformer, brief short circuits in the loads seldom cause additional damage in an SCR deflection system.

4. Since the regulator circuit operates in combination with the pincushion, width, and linearity circuits, it is good practice to make certain that the regulator section is operating normally before testing components in the associated circuits.

5. Remember that the high-voltage protection circuits operate by increasing the horizontal-oscillator frequency. In case the protection circuit should be defeated during a troubleshooting procedure, reduce the line voltage below 100 volts to avoid the development of excessive high voltage, which could damage various components.

9 . 3 TROUBLESHOOTING VERTICAL-DEFLECTION UNITS

Troubleshooting a vertical-deflection system often requires the technician to take feedback loops into consideration. Feedback results in extensive interaction of the subsections in addition to the waveshaping and amplifying functions. A typical arrangement comprises a vertical oscillator, predriver, and output stages, with a load consisting of the vertical-deflection coil windings, as shown in Fig. 9–10. Typical trouble symptoms include nonlinear deflection, subnormal picture height, abnormal picture height, keystoning, foldover, and erratic or intermittent operation.

With reference to Fig. 9–10, the vertical-output stage operates as an amplifier and as a waveshaper so that the output voltage will cause a linear sawtooth current flow through the deflection coils. Feedback from the vertical-output stage to the vertical oscillator is used to sustain oscillation. Thus, if a coupling capacitor in the feedback loop becomes "open," the vertical oscillator stops. A schematic diagram for this vertical-deflection system is shown in Fig. 9–11. Circuit action can be described by considering the vertical oscillator as a switch that closes (turns on) at retrace time, and opens (turns off) during the vertical scan time. This switching

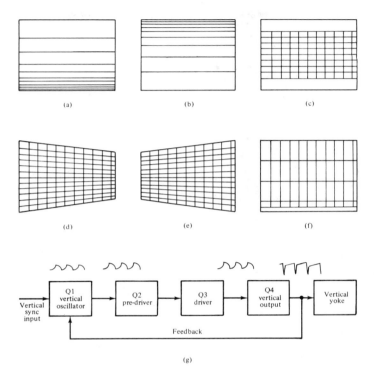

Figure 9–10. Vertical-deflection system block diagram and trouble symptoms: **(a)** Nonlinear vertical deflection; **(b)** Another example of nonlinear deflection; **(c)** Subnormal vertical height; **(d)** Vertical keystoning; **(e)** Another example of keystoning; **(f)** Abnormal vertical height; **(g)** Block diagram of vertical-sweep system.

action at the collector of the vertical-oscillator transistor functions to discharge the sweep capacitor C9 quickly to +30 volts at retrace time, and to develop a linear decrease in capacitor voltage during the scan time while the oscillator transistor is nonconducting (open). Feedback capacitor C3 is included to prevent high-frequency spurious oscillation.

If C9 becomes leaky, the picture height will be subnormal. Again, if C9 "opens," there will be a dark screen (no raster). Scanning time duration (while Q1 is "open") is determined by the time constant of C1-R8 with R14. If C1 becomes leaky, vertical sync is lost, and R14 will be thrown out of range if the leakage is substantial. Observe that sweep capacitor C9 operates in a feedback

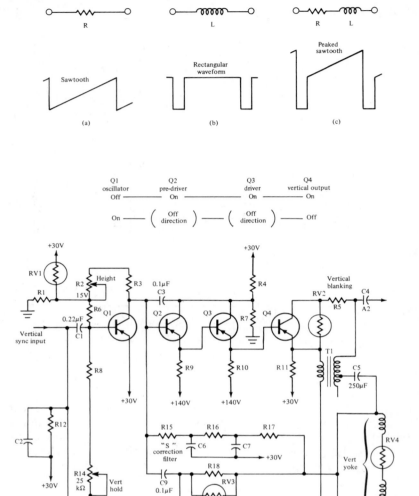

Figure 9–11. Representative vertical-deflection system: **(a)** Sawtooth voltage develops a sawtooth current through R; **(b)** Rectangular voltage develops a sawtooth current through L; **(c)** Peaked-sawtooth voltage develops a sawtooth current through series R and L; **(d)** Configuration.

loop. This is called a Miller-feedback amplifier configuration; it keeps the charging current constant so that a linear sawtooth waveform is generated. Note also that an "S" correction filter is included in this feedback loop. This is a waveshaping circuit that slows down the sweep rate slightly at the top end and at the bottom end of the scan. In other words, the iron core in the vertical-output transformer is somewhat nonlinear along its magnetic characteristic, and linear scanning requires that this nonlinearity be compensated by a suitable corrective filter. In case of nonlinearity at the top or at the bottom portion of the raster, the technician would suspect that C6 or C7 has become leaky or "open."

Observe that diode D1 and capacitor C8 in Fig. 9–11 form an antihunt circuit, to prevent amplitude variation in the flyback pulse and resulting jitter in the raster. In other words, if the flyback pulse varies in amplitude, the sawtooth amplitude will also vary. Accordingly, if C8 happens to "open," alternate cycles of the sawtooth will have different amplitudes. In turn, a vertical-jitter picture symptom occurs. Voltage-dependent resistors (varistors) are employed in this configuration to stabilize bias voltage and to clip high-amplitude pulses. In addition, varistors compensate for ambient temperature variation. Accordingly, RV2 clips the high-amplitude flyback pulse and thereby protects Q4 from breakdown damage. If Q4 is found to be defective, it is advisable to check RV2; otherwise, the replacement transistor may also fail promptly. Observe that RV1 functions to stabilize the bias voltage on Q1. This is a conventional bias-stabilization circuit. Finally, RV3 and RV4 function to maintain the picture height and linearity constant under conditions of ambient temperature variation. If a varistor is suspected of malfunction, it is advisable to make a substitution test instead of attempting to check its electrical characteristics.

REVIEW QUESTIONS

1. Briefly describe a module.
2. How many modules does a typical color-TV receiver contain?
3. Is a module necessarily complete in itself from a functional viewpoint?
4. What is the chief advantage of modular construction?

5. Explain how a "bridging" test is made in a modular receiver.

6. To what family of solid-state devices does an SCR belong?

7. Why should SCR's never be interchanged in a horizontal-deflection system?

8. Describe the result of an open-circuited deflection-coil winding in an SCR flyback system.

9. What is the function of a high-voltage runaway protective circuit?

10. How is pincushion distortion compensated in a deflection system?

11. State how the high-voltage value is related to the horizontal-deflection frequency.

12. Why is the regulator circuit checked before components in associated circuits are tested?

13. Name several trouble symptoms resulting from malfunctions in the vertical-deflection system.

14. What is the function of a voltage-dependent resistor in a vertical-sweep circuit?

15. Are varistor characteristics ordinarily checked in service shops?

Digital Color-TV Circuitry Troubleshooting

10 . 1 GENERAL CONSIDERATIONS

Troubleshooting of digital circuitry in color-TV receivers requires a basic understanding of the circuit actions that occur in normal operation. For example, a UHF/VHF tuner that employs logic (digital) circuitry operates on considerably different principles in comparison to a conventional tuner. A color receiver that utilizes this type of tuner is illustrated in Fig. 10–1. Instead of being mechanically tuned, the resonant circuits are varied by application of precise DC voltages to the tuner. These DC voltages actuate voltage-tuned circuits, as detailed subsequently. In the example of Fig. 10–1, a readout section is also provided. The readout circuit generates, positions, and controls a numerical display on the picture-tube screen. This display indicates the channel to which the receiver is tuned. When a digital-clock accessory is included, the screen display also indicates the time in a six-digit format.

Tuning is accomplished by means of digital-logic circuits that switch a series of preset voltage levels into voltage-tuned circuits that automatically select the desired channel. Digital-logic circuits consist basically of electronic switches that are called AND gates, OR gates, NAND gates, and NOR gates, supplemented by free-running and bistable multivibrators (flip-flops). It is helpful to consider these electronic-switching actions briefly. Figure

Figure 10–1. Appearance of a digital color-TV receiver. (*Courtesy of Heath Company*)

10–2 summarizes the switching characteristics of the four basic gates. When a gate terminal is "on," a DC voltage, such as +5 volts, is present. On the other hand, when a gate terminal is "off," it is at ground potential (zero volts).

If an inverter follows an AND gate, the combination functions as a NAND gate. An inverter is simply an amplifier that reverses the pulse polarity from input to output. In other words, when the input is "on," the output is "off." When the input is "off," the output is "on." Figure 10–3 shows the inverter symbol and summarizes its function. Similarly, when an inverter follows an OR gate, the combination functions as a NOR gate. Next, a flip-flop can be compared with a pushbutton switch that is connected to two lamps. Thus, when the pushbutton is depressed and then released, one of the lamps will turn "on" and the other lamp will turn "off." If the pushbutton is again depressed and released, the lamp that was "off" will now turn "on," and the lamp that was

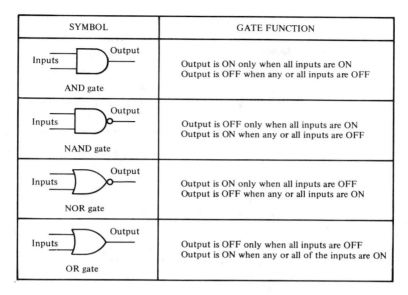

SYMBOL	GATE FUNCTION
Inputs ⟶ Output — AND gate	Output is ON only when all inputs are ON Output is OFF when any or all inputs are OFF
Inputs ⟶ Output — NAND gate	Output is OFF only when all inputs are ON Output is ON when any or all inputs are OFF
Inputs ⟶ Output — NOR gate	Output is ON only when all inputs are OFF Output is OFF when any or all inputs are ON
Inputs ⟶ Output — OR gate	Output is OFF only when all inputs are OFF Output is ON when any or all of the inputs are ON

Figure 10–2. Summary of the basic gate switching characteristics.

SYMBOL	INVERTER FUNCTION
Input ⟶ Output — Inverter	Output is ON only when the input is OFF Output is OFF only when the input is ON

Figure 10–3. Inverter symbol and function.

"on" will turn "off." Figure 10–4 shows the basic flip-flop (FF) symbol and summarizes its function. The outputs are designated Q and $\overline{Q}$; the symbol $\overline{Q}$ is called NOT Q.

Next, a latch is a flip-flop that stores "on" and "off" information (signals) until it is needed. A latch is an FF with an extra input, which is often called the latch. When the latch is "on," the information applied to the input will be transferred to the Q output. On the other hand, when the latch is "off," the information that was applied to the input at the time that the latch was turned "off" will be stored on the Q output. These functions are summarized in Fig. 10–5. Latches are basic units in counters. A counter is a series of flip-flops that are connected so that their states corre-

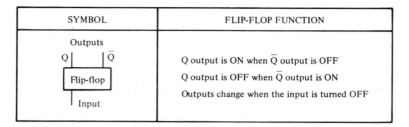

SYMBOL	FLIP-FLOP FUNCTION
Outputs Q Q̄ Flip-flop Input	Q output is ON when Q̄ output is OFF Q output is OFF when Q̄ output is ON Outputs change when the input is turned OFF

Figure 10–4. Flip-flop symbol and basic functions.

SYMBOL	LATCH FUNCTION
Output Q Q̄ Latch Latch Infor- Inputs mation	Q output will follow the information input. Q̄ output will always be the opposite of the Q output

Figure 10–5. Latch symbol and functions.

spond to the number of times that the input has been turned "on" and "off." A counter can also be utilized as a divider. Note that a digital counter indicates 1, 2, 4, 8, . . . values. As an illustration, after three input pulses have been applied, the outputs numbered 1 and 2 in Fig. 10–6 will both be turned "on." Outputs 4 and 1 will both be "on" when the count of 5 has been reached. Outputs 4, 2, and 1 are "on" when the counter indicates the number 7. These counter outputs are usually connected to a decoder network that automatically changes the 1, 2, 4, 8 system of counting to the familiar 1, 2, 3, 4 system (decimal system). Two widely used counters are the divide-by-12 and divide-by-10 types. The former type is used in digital clocks.

Next, observe the examples of basic digital circuitry shown in Fig. 10–7. In (a), two diodes provide the gate function; the transistors provide amplification and inversion. Because of inversion, a NAND output is available in addition to the AND output. Again, in (b), two diodes provide the gate function; the two transistors step up the signal level and provide inversion. Thus, both OR and NOR outputs are available. In (c), four inputs are provided; the diodes function as AND gates, and the following inverter develops

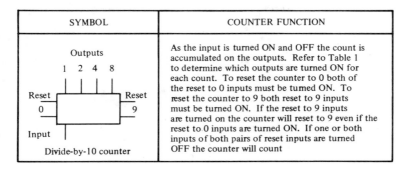

SYMBOL	COUNTER FUNCTION
  Outputs 1 2 4 8 Reset ⌐ ⌐ Reset 0 └──────┘ 9 Input ⌐ Divide-by-10 counter	As the input is turned ON and OFF the count is accumulated on the outputs. Refer to Table 1 to determine which outputs are turned ON for each count. To reset the counter to 0 both of the reset to 0 inputs must be turned ON. To reset the counter to 9 both reset to 9 inputs must be turned ON. If the reset to 9 inputs are turned on the counter will reset to 9 even if the reset to 0 inputs are turned ON. If one or both inputs of both pairs of reset inputs are turned OFF the counter will count

Figure 10–6. Divide-by-10 counter symbol and functions.

NAND-gate output. Similarly, in (d), four inputs are provided for a NOR gate; the diodes function as OR gates, and the transistor changes their outputs into a NOR level.

Two types of multivibrator arrangements are shown in Fig. 10–8. The free-running multivibrator in (a) generates a square waveform, and is generally called a "clock." Next, one type of bi-stable multivibrator is shown in (b). This is called a "toggle" configuration. When a square-wave or pulse input is applied at G, the indicator lamp will first turn "on" and then turn "off." Switching occurs on the trailing edge of the square-wave or pulse (trigger) input waveform. When bistable multivibrators (FF's) are connected in series, so that each FF triggers the following FF, a simple counter arrangement is provided. This is a 1, 2, 4, 8, . . . type of counter, as noted previously.

10 . 2   CIRCUIT FUNCTIONS IN A DIGITAL TUNER

Referring to Fig. 10–9, note the 4-bit binary up-down counter, IC202. This is a digital counter that can count from binary 0000 to binary 1111. Note that "0" means "off," and "1" means "on." The binary number 0000 corresponds to the decimal number 0, and the binary number 1111 corresponds to the decimal number 15. As seen in the diagram, the counter can produce 16 different binary numbers at its four output lines (DCBA). In other words, the counter produces various combinations in its four outputs, which are applied as inputs to the 4-line to 16-line decoder. In turn, 16 outputs are available from the decoder. These various output com-

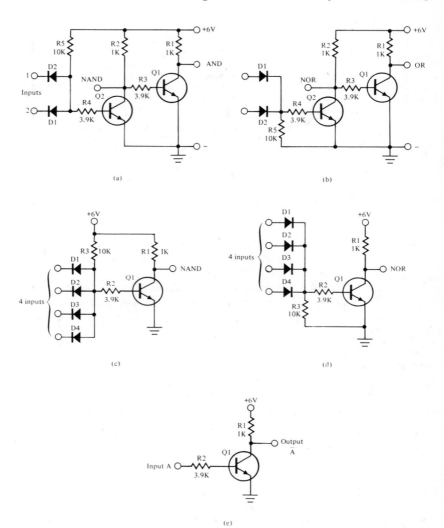

Figure 10–7. Basic digital circuitry: **(a)** Two-input configuration with AND and NAND outputs; **(b)** Two-input configuration with OR and NOR outputs; **(c)** Four-input NAND circuit; **(d)** Four-input NOR circuit; **(e)** Inverter circuit.

binations from the counter are called binary coded decimal (BCD) outputs. The combinations used in this example are summarized in Fig. 10–10.

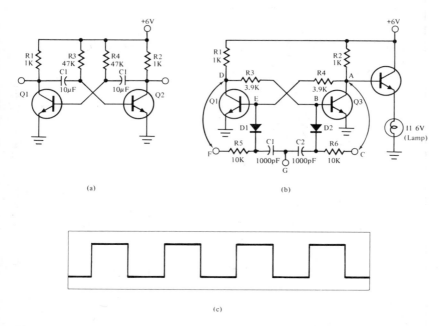

Figure 10–8. Basic multivibrator configurations: **(a)** Free-running multi-vibrator, or "clock"; **(b)** Bistable multivibrator, or "toggle"; **(c)** Waveform generated by the "clock".

A binary "up" counter consists of cascaded flip-flops that count in a 1, 2, 4, 8, . . . sequence. On the other hand, a binary "down" counter comprises cascaded flip-flops that are connected to count in a 16, 15, 14, 13, 12, . . . sequence, for example. An up-down counter is an arrangement that includes both an "up" counter and a "down" counter. When a negative-going pulse is applied to the countup input (pin 5 of IC202), the counter will count up one binary number and will then remain in that state until the next negative-going pulse is applied. Similarly, when a negative-going pulse is applied to the countdown input (pin 4 of IC202), the counter will count down one binary number and will then remain in that state until the next negative-going pulse is applied. After the counter reaches binary 1111 while counting up, it continues by starting over again at binary 0000. Also, when counting down to binary 0000, the counter continues by starting over again at binary 1111. Figure 10–11 shows a summary of the corresponding binary and decimal numbers from 0 through 15. The BCD counter

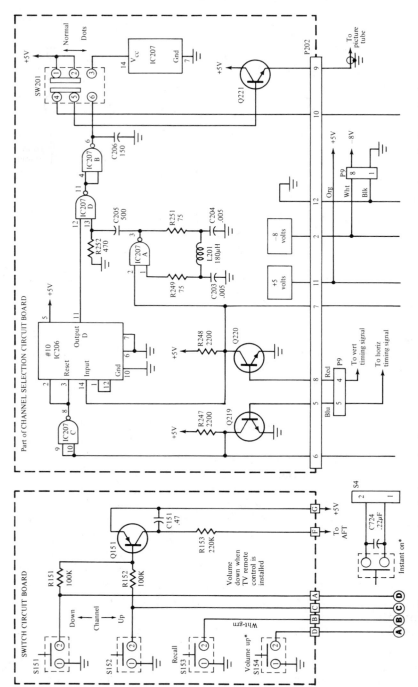

Figure 10–9. Typical channel-selection circuit-board configuration. (Courtesy of Heath Company)

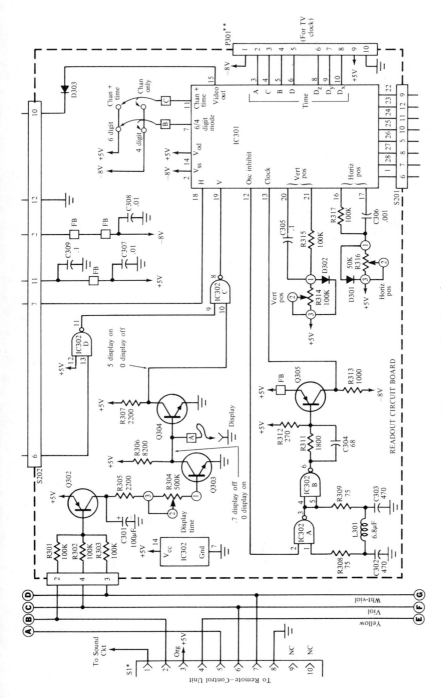

Figure 10-9. Continued

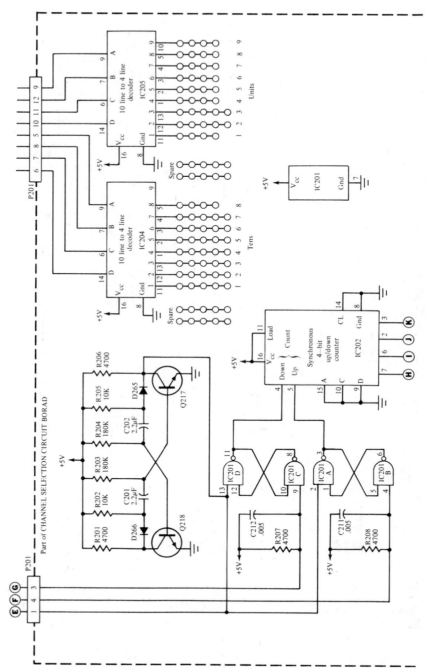

Figure 10-9. Continued

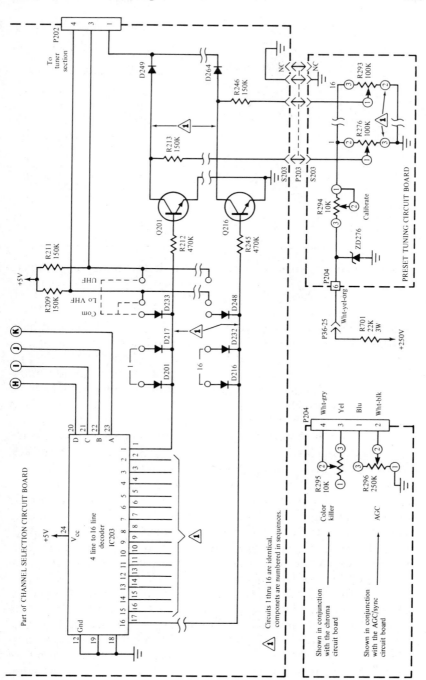

Figure 10-9. Continued

OUTPUT LINES				
D	C	B	A	
0	0	0	0	0
0	0	0	1	1
0	0	1	0	2
0	0	1	1	3
0	1	0	0	4
0	1	0	1	5
0	1	1	0	6
0	1	1	1	7
1	0	0	0	8
1	0	0	1	9
1	0	1	0	10
1	0	1	1	11
1	1	0	0	12
1	1	0	1	13
1	1	1	0	14
1	1	1	1	15

DECIMAL	BINARY
0	0
1	1
2	10
3	11
4	100
5	101
6	110
7	111
8	1000
9	1001
10	1010
11	1011
12	1100
13	1101
14	1110
15	1111

Figure 10–10. Binary coded decimal output combinations.

Figure 10–11. Corresponding binary and decimal numbers through 15.

can be regarded as functioning in a closed loop of 16 binary numbers.

Pulses required to drive the up-down counter are provided by a 2-Hz astable (free-running) multivibrator comprising Q217 and Q218. This circuit is also called the "clock." Application of these square-wave pulses to the up-down counter is controlled by IC201, a quad (four-unit) 2-input NAND-gate assembly. NAND gates C and D of IC201 form a latch for the countdown input, while gates A and B form a latch for the countup input. Normally, since one input of each of the latching gates is held high (+5 volts) by resistors R7 and R8, these latching gates prevent the 2-Hz clock waveform from entering the up-down counter. On the other hand, when the channel-up switch S152 or the channel-down switch S151 is pressed, a "low" is applied at the previously "high" input, which enables the latching gate and passes the clock signal to the respective input of the up-down counter. As long as the switch (S151 or S152) is held closed, the oscillator signal will cause the up-down counter to continue counting up or down, respectively.

Next, the 4-bit information at the outputs of the up-down counter is coupled to the four input terminals of IC203, a 4-line to 16-line decoder. Note that "bit" is an abbreviation for "binary digit." Thus, a binary "1" transition from "0" is called a "bit." One of the 16 output lines from IC203 is selected for each of the 16 binary numbers applied to its four input lines. As an illustration, when binary 0000 is present at the input of IC203, output line 1 (pin 1) of IC203 will be low (actually, it will then rest at some voltage less than 0.3 volt), and all of the other outputs (2 through 16) will be high (actually, at least 4.5 volts). These are important facts for the technician to keep in mind when troubleshooting the circuitry. If the up-down counter is advanced to 0001, then output line 2 (pin 2) of IC203 will be logic-low (less than 0.3 volt), and all other lines will be logic-high (at least 4.5 volts).

Each of the 16 output lines of IC203 is connected to a transistor-diode switch (Q201 through Q216 and D249 through D264) that switches the tuning voltage from one of the 16 tuning controls to the tuner. These tuning voltages are coupled through their respective isolation resistors to the transistor collectors and diode anodes. All of the diode cathodes are connected to the tuning-voltage terminal (4) of the tuner, as detailed subsequently. One of the 16 output lines of IC203 is low while all others are high, as noted previously. The output that is low corresponds to the binary input of IC203. The 15 outputs that are high turn on (saturate) their respective transistor switches. This places the collectors of these transistors at approximately ground potential (0.3 volt), thereby shunting the tuning voltage to ground. The output that is low causes its respective transistor to turn off and thereby permits the tuning voltage to pass through its isolation resistor and diode to the tuner. The other 15 diodes are now reverse-biased, because the positive tuning voltage appears at the cathodes of these diodes, whereas their anodes are grounded by the transistor switches. Therefore, it can be seen that the 15 saturated transistor switches "short" the 15 nonselected tuning voltages to ground, whereas they do not affect the selected tuning voltage.

Sixteen adjustable tuning voltages are obtained from controls R276 through R293. The +33-volt DC tuning supply voltage is regulated and temperature-compensated by zener diode ZD276. This voltage is then coupled through the calibrating control R294 to the 16 tuning controls. Because all tuners do not require exactly the same tuning voltage, the calibrating control is adjusted (during calibration) to set the maximum voltage across the tuning controls to correspond with the tuner requirement.

Integrated circuits IC206 and IC207 and their associated components comprise a "dot" generator for screen display. IC206 is a divide-by-10 binary counter, while IC207 is a quad 2-input NAND gate. The positive-going vertical-retrace pulses coupled to the base of Q219 turn the transistor on during the vertical-retrace period. This results in a logic 0 pulse at the collector during retrace time. This logic 0 pulse is coupled to pins 9 and 10 of IC207C. The pulse is inverted and fed out on pin 8 to pins 2 and 3 of IC206 as a logic 1. This logic 1 resets the four output lines of the divide-by-10 counter to logic 0, and also prevents it from counting while the logic 1 is present at pins 2 and 3. At the end of the vertical-retrace period, Q219 turns off and the logic 1 at pins 2 and 3 of IC206 goes to logic 0. This enables the counter to begin counting. The positive-going horizontal pulse turns Q220 on during horizontal-retrace time. In turn, a logic 0 pulse is produced during retrace and is fed to the clock input pin 14 of IC206. Therefore, the divide-by-10 counter is counting horizontal-retrace pulses, or, in effect, it is counting horizontal scan lines. Only the D output (pin 11) of the counter is being used. Since IC206 is a divide-by-10, the "D" will be logic 0 for the first eight counts, and logic 1 for the ninth and tenth counts. The D output is connected to one input (pin 12) of NAND gate IC207D, which enables it every ninth and tenth horizontal scan line.

Horizontal pulses are also fed to one input (pin 2) of NAND gate IC207A. The logic 0 disables the gate during retrace, but it is enabled during horizontal scan time when the horizontal pulse is logic 1. This gate functions as an LC oscillator with L1 and C3 and C4 providing the necessary phase shift between input and output. Resistors R49 and R51 provide isolation between input and output. The logic 1 at pin 2 of IC207A, during the horizontal scan, permits the oscillator to operate only during scan time and also synchronizes it to the horizontal-retrace pulse. The oscillator operates at a frequency of approximately 250 kHz. Resistor R252 biases pin 13 of IC207D to logic 0. The positive-going tips of the sine oscillation waveform, coupled through C205 and developed across R252, bias pin 13 to logic 1. Thus, the positive-going tips of the oscillator sine wave will enable this gate only during every ninth and tenth horizontal scan lines. The output at pin 11 is logic 0 pulses. They are coupled to pins 4 and 5 of NAND gate IC207B, inverted, and fed out of pin 6 to the dots-normal switch. The 5-volt supply is connected to IC207 only when the dots-normal switch is in the "dots" position. Transistor Q221 couples the signal, dots, or readout pulses, depending on the switch setting, to the video-output circuit board.

Next, the readout circuit generates, positions, and controls the numerical display on the picture-tube screen. The essential portion of this circuit is IC301, a CMOS (complementary metal oxide semiconductor) integrated circuit that contains most of the logic circuitry used to generate the numerical display signal. The numerical display is synchronized to the horizontal and vertical re-trace pulses. Transistors Q219 and Q220 on the channel-selecting board produce logic 0 (low) pulses during vertical and horizontal retrace times. The horizontal pulses are coupled to the horizontal sync input (pin 18) of IC301. The vertical pulses are coupled to pin 13 of IC302D, a quad 2-input NAND gate. Here, the logic 0 pulse is inverted and coupled to pin 9 of IC302C, one input of another two-input NAND gate. The other input (pin 10) is normally at logic 0, thus disabling the gate. When the channel-up, channel-down, or display-recall switches are pressed, a logic 0 is placed at pins 2, 3, or 4 of plug P1. This causes transistor Q302 to conduct and to charge capacitor C301.

Transistor Q303 is also biased on by current flowing through the display time control, R304, and resistor R305. When the activated switch is released, Q302 ceases to conduct. Transistor Q303, how-ever, remains turned on, owing to the positive charge remaining on C301, until C301 can no longer supply enough current to keep it con-ducting. The discharge time of C301 controls the length of time that the number will be displayed on the picture-tube screen after the command signal has been removed. When transistor Q303 is on, Q304 is biased off and places a logic 1 at one input (pin 10) of NAND gate IC302C. This enables the gate and allows vertical pulses to pass through it to the vertical-sync input (pin 19) of IC301. The display is enabled continuously when the jumper wire at A is connected to the display pin. This keeps Q304 turned off, enabling NAND gate IC302C. Timing circuits consisting of R314, C305, R316, and C306 control the vertical and horizontal positions of the numerical display on the screen.

NAND gates I302A and IC302B, transistor Q305, and the associ-ated circuitry form a gated oscillator and driver circuit. The oscillator frequency is determined by coil L301 and capacitors C302 and C303. When the input (pin 2) of NAND gate IC302A is at logic 1, the gate is enabled and the circuit oscillates. The logic 1 appears at the input of this gate only when both vertical and horizontal sync signals are present at IC301. The output signal of IC302A is coupled to NAND gate IC302B, where it is inverted and "squared up." These pulses are then coupled through resistor R311 and capacitor C304 to the base of transistor Q305. The signal that appears at the collector of Q305 is a square wave that varies from +5 volts (logic 1) to −9 volts (logic 0) to match the logic levels required by the readout IC. This gated oscilla-tor signal is coupled to the clock input pin (pin 13) of IC301.

Channel number data are coupled to IC301 from IC204 and IC205 on the channel-selecting circuit board. This information is processed within the readout IC to produce the video-output signal. "Channel only" or "channel and time" (when the clock accessory is installed) is selected by a jumper wire at point C. A "4-digit" or "6-digit" clock display is selected by a jumper wire at point B. The UHF/VHF tuner electronically selects the desired television signal from channels 2 through 83. A tuning voltage is supplied by the channel-selecting circuit board. Figure 10–12 shows a block diagram of the UHF/VHF tuner.

A balanced 300-ohm UHF antenna input connects to a voltage-tuned circuit A, which provides selectivity prior to amplification by amplifier B, a common-base bipolar stage. Output from this amplifier is coupled through another voltage-tuned circuit C, to an autodyne converter (mixer D). Here, the signal is combined with oscillator signal E, to produce an intermediate frequency in the 45-MHz range. This signal is then coupled through a matching network F, to a bipolar amplifying stage L. This stage functions as a 45-MHz amplifier when receiving UHF signals and as a cascode mixer for VHF signals. A broadly tuned circuit G couples the IF output signal from the tuner to the IF amplifier. The VHF portion of the tuner employs a 75-ohm antenna input. However, balun transformer T703 converts a 300-ohm VHF antenna input to the required 75-ohm impedance.

A voltage-tuned circuit H provides selection of the VHF signal before it is amplified by a diode-protected dual gate MOSFET circuit J. The amplifier is coupled through a double-tuned coupling and matching network K, to a cascode mixer stage L. Here the selected and amplified VHF signals are heterodyned by the signal from the oscillator circuit M. The same broadly tuned circuit G that couples the UHF signal to the IF-amplifier circuit also couples the VHF signal to the IF amplifier. When the tuner is programmed, an appropriate B+ voltage (UHF B+ or VHF B+) is applied to the tuner from the U-V switching circuit board, shown in Fig. 10–13. Also, separate voltages are used to program the LO VHF (channels 2–6), and HI VHF (channels 7–13) portions of the tuner. Approximately −15 volts is used for LO VHF channels, while approximately +20 volts is supplied for HI VHF channels.

Next, consider the U-V (UHF-VHF) switching circuit that controls the application of control and supply voltages to the UHF and VHF sections of the tuner. The basic U-V switching is controlled or determined by the presence or absence of "jumper" wires used on the channel-selecting circuit board. These jumpers

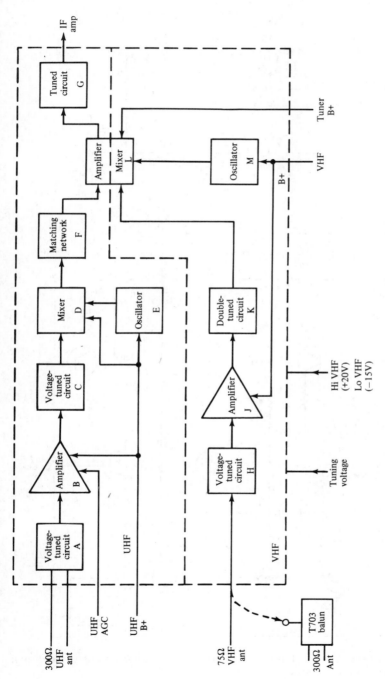

Figure 10–12. UHF/VHF tuner block diagram. (Redrawn by permission of Heath Company)

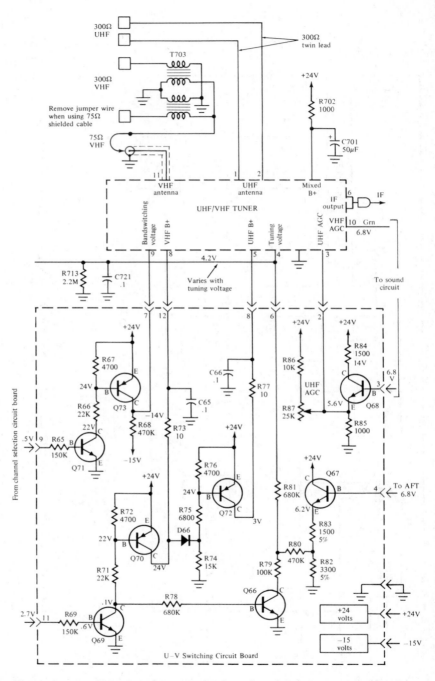

Figure 10–13. U-V switching-circuit board and tuner circuitry. (Redrawn by permission of Heath Company)

provide inputs to the circuits on the U-V switching circuit board that apply B+ voltage to the appropriate section (UHF-VHF) of the tuner, as well as switching the VHF section between the low VHF (channels 2–6) and high VHF (channels 7–13) circuitry. Transistors Q70 and Q72 are switches that supply +24 volts B+ to the UHF and VHF sections of the tuner. However, only one of these sections has power applied at a particular time. Normally, when no jumpers are installed, transistor Q69 is biased on by the positive voltage applied to its base through R69 and R211 (on the channel-selecting circuit board). When transistor Q69 is on, it biases Q70 on, which applies +24 B+ volts through decoupling network R73 and C65 to the VHF section of the tuner. Diode D66 is now forward-biased and places the junction of R74 and R75 at slightly less than 24 volts. Current through R75 and R76 is insufficient to turn on Q72. This removes +24 volts B+ from the UHF section of the tuner.

When a jumper is installed between the common and UHF connectors (on the channel-selecting circuit board), the junction of R69 and R11 will be low when the programming forward-biases the diode associated with the common connector. A logic 0 (low) is present on the common connector of the selected tuning position. This turns Q69 off, which in turn biases Q70 off and removes +24 volts B+ from the VHF section of the tuner. Diode D66 is now forward-biased, and the voltage divider consisting of R74, R75, and R76 biases Q72 on. When Q72 is on, it applies +24 volts B+ through decoupling network R77 and C66 to the UHF section of the tuner.

A tuning-error correction voltage from the AFT circuit board is applied to the base of emitter-follower Q67. This error-correction voltage is divided by resistors R82 and R83, and then coupled through R80 and R81 to the tuning voltage input of the tuner. When a UHF channel is selected, Q66 turns on and brings one end of R79 to ground potential. The correction voltage is then divided again by R79 and R80 before it is coupled through R81 to the tuner. Because the VHF section of the tuner must cover two widely separated frequency bands (channels 2–6 from 54 to 88 MHz, and channels 7–13 from 174 to 216 MHz), the amount of inductance and capacitance in the tuning circuits is switched. Diodes in the tuner are used to electrically switch the required values of inductance and capacitance for each band into the circuit. These diodes are controlled by the band-switching voltage, which is −15 volts for the low VHF channels, and +20 volts for the high VHF channels.

When no jumpers are installed, Q71 is biased on by the positive voltage applied to its base through R65 and R209 (on the channel-selecting circuit board). Transistor Q71 then biases Q73 on and places +24 volts at the band-switching input of the tuner. This switches the tuner to the high VHF band. When a jumper is installed between the common and low VHF connectors (on the channel-selecting circuit board), the junction of R65 and R209 will be low when the programming forward-biases the diode associated with the common connector. This turns Q71 off, which in turn biases Q73 off, and places −15 volts at the band switching input of the tuner. This switches the tuner to the low VHF band. AGC voltage is coupled to the base of Q68, which is an emitter follower. This voltage varies from approximately +10 volts with no incoming signal to about 0.5 volt with a very strong incoming signal. The UHF AGC control R87 sets the maximum AGC voltage (approximately 1.2 volts) to match the maximum gain reduction point of the UHF tuner. Thus, even when very strong incoming signals are processed, and Q68 turns off, the AGC voltage to the UHF tuner does not drop below the value set by the UHF AGC control. VHF AGC is applied directly to the tuner.

10 . 3   DIGITAL TEST EQUIPMENT

In addition to conventional test equipment such as the TVM and oscilloscope, specialized digital test equipment is often helpful. For example, when incorrect readout occurs, the questions that arise concern whether a gate is functioning, whether a pin of an IC might be short-circuited to ground or short-circuited to the supply voltage, or whether a counter is counting. If suitable test instruments are used, the questions can be answered without unsoldering IC pins, or cutting circuit-board conductors. For example, a logic probe, such as that illustrated in Fig. 10–14, is very useful, and is considered by numerous experienced technicians as the single most important test instrument for troubleshooting digital equipment. This probe is used to trace logic levels and pulses through integrated circuitry to determine whether the point under test is logic-high, logic-low, bad level, open-circuited, or pulsing. This probe has preset logic threshold levels of 2.0 volts (logic-high), and 0.8 (logic-low) volt. When the probe is applied at a logic-high terminal, a bright band of light appears around the probe tip. On the other hand, when the probe is touched to a

Figure 10–14. A logic probe for troubleshooting digital circuitry. *(Courtesy of Hewlett-Packard, Inc.)*

logic-low terminal, the light goes out. Open circuits or "bad-level" voltages produce a band of light at half brightness; the lamp flashes on or blinks off, depending on the polarity of the pulse. Pulse trains with frequencies up to 50 MHz cause the lamp to blink on and off at a 10-Hz rate.

To check whether pulse trains may be missing, the logic probe can be applied at various test points while the circuit is operated at normal speed. For a more exacting test, the circuit can be stepped one pulse at a time while the output response is checked. This type of test requires a single-shot pulse generator. For example, a logic pulser such as that illustrated in Fig. 10–15 is generally used for this purpose. Probe and pulser operation is automatic, and no adjustments are required. Thus, the technician is free to turn his attention to circuit analysis instead of being distracted by instrument adjustments. Connectors used with the logic

Figure 10–15. A logic pulser is basically a single-shot pulse generator. *(Courtesy of Hewlett-Packard, Inc.)*

probe and logic pulser provide a connecting lead to a 5-volt power supply, either in the circuit under test or from an external power supply. A ground clip is provided to connect the probe return lead to the ground or common bus of the circuit under test.

Figure 10–16 shows how a logic pulser and logic probe are applied to check the response of an IC. The pulser is applied to the input test terminal, and the logic probe is applied at an output terminal. In turn, the pulse button is pressed on the logic pulser, and the logic probe indicates the resulting response (or failure to respond). When the pulser button is pressed, low nodes are automatically pulsed high; or, if the node is high, it is automatically pulsed low. Logic pulsers and probes are available with various threshold levels, to accommodate different types of logic circuitry. Another useful digital-circuit test instrument is called the *logic clip*, illustrated in Fig. 10–17. This tester clips over an IC, and makes connection to all of the terminals. It instantly displays the logic states of all 14 or 16 terminals. Sixteen light-emitting diodes (LED's) in the tester independently follow level changes at each IC terminal. An illuminated LED indicates a logic-high state. It is used with good advantage with a logic pulser. Defective IC's become immediately apparent, because a defective IC will not go through its specified sequence of logic-high and logic-low states. Sequences of logic states indicated by the logic clip are compared with the sequences specified in the receiver service data.

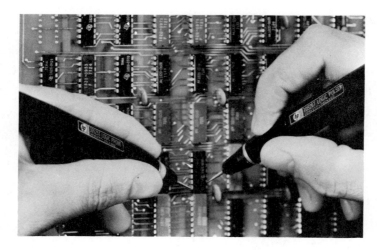

Figure 10–16. Logic pulser and logic probe applied to check IC action. *(Courtesy of Hewlett-Packard, Inc.)*

Figure 10–17. Logic clip shows DTL and TTL logic states at a glance. *(Courtesy of Hewlett-Packard, Inc.)*

REVIEW QUESTIONS

1. How is tuning accomplished by digital-logic circuits?
2. Explain the operation of an AND gate.
3. What is the function of a latch?
4. State the binary counting sequence.
5. Distinguish between a free-running and a bistable multivibrator.
6. How does an "up" counter differ from a "down" counter?
7. Define a binary "bit."
8. Compare OR-gate operation with AND-gate operation.
9. How do NAND and NOR gates differ from AND and OR gates?
10. Can diodes be used as switches?
11. Briefly describe a logic probe.
12. Name typical logic-high and logic-low threshold values.
13. What is the function of a logic pulser?
14. Do all types of logic circuitry employ the same high and low values?
15. Briefly discuss the function of a logic clip.

Answers to Review Questions

CHAPTER 1

1. The basic troubleshooting approach starts with an analysis of the picture and sound symptoms.

2. More than one kind of fault may be responsible for a particular trouble symptom.

3. To reconstitute the color subcarrier, the color burst is processed for locking the subcarrier oscillator.

4. Color sync may be lost without concurrent loss of horizontal sync.

5. When horizontal sync is lost, color sync is necessarily lost also.

6. If the front end has deficient frequency response, it may reject the color portion of the incoming composite color signal.

7. Confetti is colored snow displayed on the picture-tube screen.

8. Convergence denotes a picture-tube operating condition wherein all three beams focus at the same point on the screen.

9. A logical troubleshooting procedure starts with a discussion of the onset and nature of the trouble symptoms with the set-owner.

10. A hi-lo multimeter has all the functions of a conventional multimeter, plus a low-voltage ohmmeter function.

11. An oscilloscope used in color-TV service needs to have a vertical-amplifier frequency response that is uniform through 3.58 MHz.

12. Signal outputs provided by a typical keyed-rainbow and pattern generator are a keyed-rainbow signal, white-dot signal, and crosshatch signal.

13. The chief applications for a sweep-frequency and marker generator are in wide-band alignment procedures for RF, IF, and chroma circuitry.

14. Tube testers are employed in present-day color-TV service shops.

15. A color picture-tube test jig provides a quick check to determine whether a trouble symptom is being caused by a defective picture tube or by a circuit malfunction.

CHAPTER 2

1. A color-TV receiver has all the sections utilized in a black-and-white receiver, plus a chroma section with a color picture tube.

2. The color-subcarrier frequency is 3.58 MHz, approximately.

3. Horizontal sync pulses have significant frequencies extending to approximately 1.6 MHz.

4. Capacitor failure (an open-circuit condition) can cause misalignment of an RF or IF tuned circuit.

5. High-frequency response in a video amplifier is determined by the values of the collector load resistors and inductances of the peaking coils.

6. Color-sync action is impaired by a mistimed gating pulse because the color burst becomes attenuated or suppressed.

7. An oscilloscope is used to check the amplitudes and shapes of chroma waveforms in analysis of color-sync circuit operation.

8. IF regeneration causes distortion of the frequency response curve, often with development of sharp high-amplitude peaks at one or more frequencies.

9. Hue is a function of chroma phase with respect to the color-burst phase.

10. Chroma phases are usually checked with a keyed-rainbow signal and display of the chroma-demodulator output waveforms on a scope screen.

11. A common cause of IF overload is AGC malfunction.

12. The chief features of a television analyzer are signal outputs of various frequencies and characteristics for signal-substitution tests in the RF, IF, video, chroma, sync, and sweep sections of a TV receiver.

13. An NTSC color-bar generator differs from a keyed-rainbow generator in that the former provides primary and complementary colors at full saturation and brightness.

14. Almost any trouble symptom in a tube-type receiver throws initial suspicion upon the tubes.

15. A tube tester generally checks tubes for mutual conductance and/or emission, and interelectrode short-circuits. A gas test may also be provided.

CHAPTER 3

1. Color-killer control voltages enable and disable the bandpass amplifier.

2. The chief function of a bandpass amplifier is to separate the chroma signal from the Y signal.

3. Color "fit" denotes the accuracy with which the color image overlays the black-and-white image.

4. A color-killer control is set correctly when set just below the point that the color image will be suppressed during reception of a weak color signal.

5. Localization of leaky capacitors can often be accomplished on the basis of DC-voltage measurements. In some cases, supplementary data can be obtained by making transistor turn-off and/or turn-on tests.

6. When DC-voltage values are specified in receiver servicing data both for color reception and for black-and-white reception, useful trouble clues are provided by measurements under both signal conditions.

7. In-circuit resistance-measurement procedures are often made to advantage with a hi-lo ohmmeter. Alternatively, PC conductors can be temporarily open-circuited.

8. A transistor turn-off test shows whether collector current is cut off when base and emitter are brought to the same potential. A turn-on test shows whether collector current increases when the forward bias is increased.

9. Turn-off or turn-on tests may be particularly informative when the troubleshooter is not certain whether there is a defect in a transistor or an associated component.

10. Tube-type chroma circuitry differs from solid-state chroma circuitry chiefly in the higher DC voltages that are employed in the former.

11. Chroma-circuit malfunction is generally assumed to be caused by faults other than misalignment.

12. Trouble in the ACC section can cause abnormally high color intensity, subnormal color intensity, drifting, and intermittent color reproduction.

13. An oscilloscope is often helpful in ACC troubleshooting procedures to check the amplitudes and shapes of pertinent waveforms.

14. Chroma waveform analysis is concerned with amplitude, waveshape, frequency, and phase relations.

15. Many waveshapes have a comparatively wide tolerance on width; some, such as chroma-demodulator output waveforms, have a fairly tight tolerance on phase; most waveforms have a ±20% tolerance on amplitude.

CHAPTER 4

1. Two basic types of subcarrier-oscillator configurations are the ringing-crystal arrangement and the crystal-controlled free-running oscillator with APC control.

2. A regenerated subcarrier is a locally-generated 3.58-MHz voltage that is phase-locked to the color burst.

3. The frequency of a subcarrier-oscillator crystal is controlled by the capacitance variation provided by a varactor diode.

4. Burst amplifiers have less bandwidth than bandpass amplifiers.

5. Analysis of color-sync trouble symptoms usually starts with DC-voltage measurements.

6. The peak frequency of the burst-amplifier response curve is 3.58 MHz.

7. A color-TV transmission employs a subcarrier frequency of 3.58 MHz, approximately, whereas a keyed-rainbow signal has a subcarrier frequency of approximately 3.56 MHz.

8. More diagonal color "rainbows" are displayed on the picture-tube screen as the color-subcarrier oscillator drifts farther off-frequency.

9. Output from a signal generator can be used to make a substitution test in the subcarrier-oscillator circuit.

10. If receiver service data are unavailable, the troubleshooter may be able to make comparison tests and measurements on a similar receiver that is in good operating condition.

11. A module is similar to a circuit board, with a plug-in connected at one end.

12. Variable resistors are more likely to deteriorate than are fixed resistors.

13. A lab-type signal generator is provided with a carrier-level meter and a calibrated output attenuator, whereas the output level from a service-type signal generator is uncalibrated.

14. An RC substitution box is useful to make quick checks of circuit response to various values of resistance or capacitance.

15. A vectorgram depicts demodulation phases, demodulation linearity, relative signal amplitudes, and various forms of distortion.

CHAPTER 5

1. A chroma demodulator decodes the chroma signal into R-Y, B-Y, and G-Y components.

2. R-Y and B-Y signals are separated in a chroma demodulator by phase-detection with reference to the injected subcarrier voltage.

3. When the hue control is turned, a vectorgram pattern will rotate on the oscilloscope screen.

4. Capacitor defects are often responsible for incorrect demodulation phase angles.

5. The resistance value of an inductor is unrelated to its inductance value.

6. An oscilloscope is useful in analysis of demodulator trouble symptoms for input/output waveform checks, phase measurements, and linearity tests.

7. Demodulator output waveforms are checked for correct phases on the basis of a keyed-rainbow signal, and display of demodulator output waveforms on an oscilloscope screen. Vectorgrams may be utilized for this purpose.

8. ATC action can simulate chroma-demodulator malfunction if the technician is unaware of the waveform modification that is normally introduced.

9. An R-Y/B-Y demodulation system is a quadrature system, whereas an XZ system operates at some other angle, such as 120 deg.

10. When malfunction occurs in a tube-type chroma-demodulation circuit, tubes are checked first.

11. Vectorgrams would have much higher harmonic content and would have a different waveshape if chroma demodulators had the same bandwidth as video amplifiers.

12. If a chroma demodulator develops nonlinear output, the associated semiconductor device(s) falls under initial suspicion.

13. Chroma-waveform baseline curvature affects the central aspect of a vectorgram display.

14. Horizontal blanking pulses may not appear in vectorgram displays, depending upon details of bandpass- and burst-amplifier circuitry.

15. A blanking "petal" can be quickly distinguished from a vectorgram "petal" by turning the color control. A blanking "petal" will be unaffected, whereas a vectorgram "petal" will change amplitude.

CHAPTER 6

1. A chroma matrix combines two chroma signals to develop a third chroma signal.

2. An oscilloscope may be employed to check the input/output waveform relations in a matrix arrangement, to measure peak-to-peak voltages, and to check waveshapes.

3. A G-Y signal can be matrixed from X and Z signals.

4. To disconnect a component from a circuit board without desoldering, one or more PC conductors may be temporarily cut.

5. A cold-soldered connection can simulate high resistance, an open-circuit condition, or intermittent operation.

6. An RGB matrix system combines chroma signals and a Y signal to form red, green, and blue color signals.

7. To operate a color picture tube as an RGB matrix, the Y signal is applied to the picture-tube cathodes, and the R-Y, B-Y, and G-Y signals are applied to the grids of the tube.

8. An oscilloscope is often useful in analyzing matrix malfunctions by showing points of signal loss, attenuation, or distortion, and amplitude relations.

9. In pre-demodulator matrixing, the Y signal is combined with the chroma signal before demodulation, or simultaneously with demodulation. In post-demodulator matrixing, the Y signal is combined with the demodulated chroma signal.

10. Matrix output signal proportions are not the same in all receivers.

11. An NTSC color-bar generator has an advantage in checking RGB matrix action because the generator provides a Y signal in addition to chroma signals.

12. An NTSC-signal vectorgram consists essentially of "dots", whereas a keyed-rainbow vectorgram consists basically of "petals".

13. The color-bar sequence is always the same in a keyed-rainbow pattern.

14. An NTSC color-bar pattern may have an arbitrary color sequence.

15. G-Y and G-Y $\underline{/90°}$ outputs are quadrature signals.

CHAPTER 7

1. A television analyzer is used to localize faults in the signal channel by injection of suitable test signals at appropriate circuit points.

2. Four signals considered in a flow chart are the chroma signal, the Y signal, the sound signal, and the sync signal.

3. A picture-but-no-sound trouble symptom indicates a fault in the intercarrier-sound channel or in the audio section.

4. A raster may be visible, although there is no picture reproduction; however, if there is no raster present, a picture signal will not be reproduced.

5. Signal-tracing tests may be difficult or impractical in low-level circuits such as the RF section.

6. A signal-injection test can always be made in the picture channel.

7. Statistically, the horizontal-deflection and high-voltage section develops malfunctions more often than the other receiver sections.

8. A picture signal can always be disclosed by an oscilloscope waveform check.

9. Incorrect voltages at a color picture-tube socket can simulate a defective picture tube.

10. A color picture-tube test jig permits the technician to quickly distinguish between picture-tube trouble and receiver-circuit trouble.

11. It is not possible for high-voltage output to be present in the absence of horizontal-scanning action.

12. High voltage can be supplied to a picture tube from a television analyzer.

13. A systematic approach to an obscure malfunction avoids the "shotgun" method wherein much time is often wasted in random replacement of components.

14. A flyback transformer can be checked with a ringing test provided by a television analyzer, or a "ringer" accessory for an oscilloscope.

15. A deflection yoke can be checked with a ringing test provided by a television analyzer, or a "ringer" accessory for an oscilloscope.

CHAPTER 8

1. Color picture-tube circuitry can be subdivided into signal, bias, and convergence sections.

2. Two principal types of color picture tubes are the shadow-mask type and the aperture-grille type.

3. Dynamic convergence serves to bring all three beams into focus simultaneously at any point on the picture-tube screen.

4. A wide blue-field adjustment brings the vertical height of the blue field the same as that of the red field and the green field.

5. Pincushion distortion appears as curvature at the top and bottom and/or sides of the raster.

6. A parabolic dynamic-convergence waveform is developed by partial integration of the sawtooth deflection waveform.

7. A conventional dynamic-convergence system provides 13 convergence controls.

8. Many dynamic-convergence controls tend to interact because the control fields are not confined entirely to their relevant guns.

9. A shadow-mask picture tube employs phosphor color dots on its screen, whereas an aperture-grille picture tube utilizes vertical stripes of color phosphors.

10. Aperture-grille picture tubes and shadow-mask picture tubes are driven by the same form of video signal.

11. An aperture-grille picture tube cannot be used as an RGB matrix.

12. An ITR differs from an SCR in that the former contains an SCR and a diode in the same housing.

13. A degaussing coil serves to demagnetize a color picture tube and its supporting structure.

14. A thermistor operates to limit the time of current flow through a built-in degaussing-coil circuit.

15. An instant-on type of receiver should not be operated from a switched wall outlet because the picture tube could be damaged from application of high voltage before the cathodes have warmed up adequately.

CHAPTER 9

1. A module is similar to a circuit board except that it has plug-in facilities at one end.

2. Nine modules are utilized in a typical receiver.

3. A module is not necessarily complete in itself from a functional viewpoint.

4. The chief advantage of modular construction is a great reduction in the "down time" that a receiver is out of service.

5. To make a "bridging" test in a modular receiver, a small capacitor is temporarily connected from the tuner output to the 4.5-MHz sound-input terminal.

6. An SCR belongs to the same family of solid-state devices as the triac, a gate-controlled full-wave silicon switch.

7. SCR's should never be interchanged in a horizontal-deflection system because their application characteristics vary from one circuit to another.

8. An open-circuited deflection-coil winding in an SCR flyback system is very likely to result in destruction of the SCR.

9. High-voltage runaway protective circuits prevent the possibility of excessive high-voltage output, with resulting component damage and X radiation.

10. Pincushion distortion is compensated in a deflection system by introduction of waveform modifications.

11. The high-voltage value is maximum at an operating frequency of 15,750 Hz; as the scanning frequency increases, the high-voltage value decreases.

12. It is advisable to check the regulator circuit before components in associated circuits are tested, because of the possibility of simulated trouble symptoms.

13. Typical vertical-sweep trouble symptoms include nonlinear deflection, subnormal picture height, abnormal picture height, keystoning, foldover, and erratic operation.

14. A voltage-dependent resistor in a vertical-sweep circuit functions to maintain the raster height constant under varying temperature conditions.

15. Varistor characteristics are seldom checked in service shops; a substitution test is ordinarily made.

CHAPTER 10

1. Tuning is accomplished in digital-logic circuits by application of suitable bias voltages to varactor diodes.

2. An AND gate is an electronic switch that develops an output provided that all of its inputs are energized simultaneously.

3. A latch functions as the electronic equivalent of a toggle switch.

4. The binary counting sequence is 0, 1, 10, 11, 100, and so on.

5. Free-running multivibrators oscillate continuously, whereas a bistable multivibrator changes state only when it is triggered.

6. An "up" counter starts at zero and counts up to some predetermined limit; a "down" counter starts at a predetermined limit and counts down to zero.

7. A binary bit is defined as a binary digit, or, as a voltage transition from high to low (or from low to high).

8. An OR gate will produce an output whenever one of its inputs is energized.

9. NAND and NOR gates have oppositely polarized output voltages, compared to AND and OR gates.

10. Diodes can be, and often are, used as switches in electronic circuitry.

11. A logic probe provides a visible indication of a logic-high condition, a logic-low condition, or the presence of a pulse train.

12. Typical logic-high and logic-low threshold values are 2.0 volts and 0.8 volt, respectively.

13. A logic pulser is a small single-shot pulse generator.

14. Various types of logic circuitry employ different high and low values.

15. A logic clip is a test device that clips over an integrated circuit and provides a visual indication of the high and low states at its terminals.

Index